Advances in Anatomy, Embryology and Cell Biology
Ergebnisse der Anatomie und Entwicklungsgeschichte
Revues d'anatomie et de morphologie expérimentale

Springer-Verlag Berlin · Heidelberg · New York

This journal publishes reviews and critical articles covering the entire field of normal anatomy (cytology, histology, cyto- and histochemistry, electron microscopy, macroscopy, experimental morphology and embryology and comparative anatomy). Papers dealing with anthropology and clinical morphology will also be accepted with the aim of encouraging co-operation between anatomy and related disciplines.

Papers, which may be in English, German or French, are normally commissioned, but origina papers and communications may be submitted and will be considered so long as they deal with a subject comprehensively and meet the requirements of the "Advances".

For speed of publication and breadth of distribution, this journal appears in single issues which can be purchased separately; 6 issues constitute one volume.

It is a fundamental condition that submitted manuscripts have not been, and will not simultaneously be submitted or published elsewhere. With the acceptance of a manuscript for publication, the publisher acquire full and exclusive copyright for all languages and countries.

25 copies of each paper are supplied free of charge.

Die Ergebnisse dienen der Veröffentlichung zusammenfassender und kritischer Artikel aus dem Gesamtgebiet der normalen Anatomie (Cytologie, Histologie, Cyto- und Histochemie, Elektronenmikroskopie, Makroskopie, experimentelle Morphologie und Embryologie und vergleichende Anatomie). Aufgenommen werden ferner Arbeiten anthropologischen und morphologisch-klinischen Inhalts, mit dem Ziel, die Zusammenarbeit zwischen Anatomie und Nachbardisziplinen zu fördern.

Zur Veröffentlichung gelangen in erster Linie angeforderte Manuskripte, jedoch werden auch eingesandte Arbeiten und Originalmitteilungen berücksichtigt, sofern sie ein Gebiet umfassend abhandeln und den Anforderungen der „Ergebnisse" genügen. Die Veröffentlichungen erfolgen in englischer, deutscher und französischer Sprache.

Die Arbeiten erscheinen im Interesse einer raschen Veröffentlichung und einer weiten Verbreitung als einzeln berechnete Hefte; je 6 Hefte bilden einen Band.

Grundsätzlich dürfen nur Arbeiten eingesandt werden, die nicht gleichzeitig an anderer Stelle zur Veröffentlichung eingereicht oder bereits veröffentlicht worden sind. Der Autor verpflichtet sich, seinen Beitrag auch nachträglich nicht an anderer Stelle zu publizieren.

Die Mitarbeiter erhalten von ihren Arbeiten zusammen 25 Freiexemplare.

Les résultats publient des sommaires et des articles critiques concernant l'ensemble du domaine de l'anatomie normale (cytologie, histologie, cyto- et histochimie, microscopie électronique, macroscopie, morphologie expérimentale, embryologie et anatomie comparée). Seront publiés en outre les articles traitant de l'anthropologie et de la morphologie clinique, en vue d'encourager la collaboration entre l'anatomie et les disciplines voisines.

Seront publiés en priorité les articles expressément demandés, nous tiendrons toutefois compte des articles qui nous seront envoyés dans la mesure où ils traitent d'un sejet dans son ensemble et correspondent aux standards des «Revues». Les publications seront faites en langues anglaise, allemande et française.

Dans l'intérêt d'une publication rapide et d'une large diffusion les travaux publiés paraitront dans des cahiers individuels, diffusés séparément: 6 cahiers forment un volume.

En principe, seuls les manuscrits qui n'ont encore été publiés ni dans le pays d'origine ni à l'éntranger peuvent nous être soumis. L'auteur s'engage en outre à ne pas les publier ailleurs ultérieurement.

Les auteurs recevront 25 exemplaires gratuits de leur publication.

Manuscripts should be addressed to/Manuskripte sind zu senden an/Envoyer les manuscrits à:

Prof. Dr. A. **BRODAL**, Universitetet i Oslo, Anatomisk Institutt, Karl Johans Gate 47 (Domus Media), Oslo 1/Norwegen

Prof. W. **HILD**, Department of Anatomy, Medical Branch, The University of Texas, Galveston, Texas 77550/USA

Prof. Dr. J. van **LIMBORGH**, Universiteit van Amsterdam, Anatomisch-Embryologisch Laboratorium, Mauritskade 61, Amsterdam-O/Holland

Prof. Dr. R. **ORTMANN**, Anatomisches Institut der Universität, Lindenburg, D-5000 Köln-Lindenthal

Prof. Dr. T. H. **SCHIEBLER**, Anatomisches Institut der Universität, Koellikerstraße 6, D-8700 Würzburg

Prof. Dr. G. **TÖNDURY**, Direktion der Anatomie, Gloriastraße 19, CH-8006 Zürich/Schweiz

Prof. Dr. E. **WOLFF**, Collège de France, Laboratoire d'Embryologie Expérimentale, 49 Avenue de la belle Gabrielle, Nogent-sur-Marne 94/Frankreich

Advances in Anatomy, Embryology and Cell Biology
Ergebnisse der Anatomie und Entwicklungsgeschichte
Revues d'anatomie et de morphologie expérimentale

53/2

Peter Kaufmann and Michail Davidoff

The Guinea-Pig Placenta

With 21 Figures

Springer-Verlag Berlin · Heidelberg · New York 1977

Priv.-Doz. Dr. Peter Kaufmann, Anatomical Institute, University of Hamburg, Martini-straße 52, D - 2000 Hamburg 20, Federal Republic of Germany

Doz. Dr. Michail Davidoff, Regeneration Research Laboratory, Bulgarian Academy of Sciences, 1431 Sofia, Bulgaria

ISBN-13: 978-3-540-08179-1   e-ISBN-13: 978-3-642-66618-6
DOI: 10.1007/ 978-3-642-66618-6

Library of Congress Cataloging in Publication Data. Kaufmann, Peter, 1942-   The guinea-pig pla-centa. (Advances in anatomy, embryology, and cell biology; v. 53, fasc. 2) Bibliography: p. Includes index. 1. Placenta. 2. Guinea-pigs-Anatomy. 3. Guinea-pigs-Physiology. I. Davidoff, Michail S., joint author. II. Title. III. Series. QL801.E67 vol. 53, fasc. 2 [QL881] 574.4'08s [599'.3234] 77-2842

Composition: H. Stürtz AG, Universitätsdruckerei, Würzburg
2121/3321-543210

Contents

1. Introduction

The placenta of the guinea-pig has assumed exceptional importance among the discoidal hemochorial placentae since the end of the sixties. Up to that date, numerous studies had been published on the functional analysis of the human placenta. One shortcoming common to all these studies, however, was the fact, that the human placenta was not absolutely suitable for morphological research, owing to fixation difficulties, and for experimental investigations virtually unsuitable. Since other hemochorial villous placentae, like that of the anthropoid apes (primates) were practically unobtainable, a possible alternative was sought among the lacunal placentae. The numerous studies on the guinea-pig placenta undertaken at the turn of the century favoured the choice of this organ as it also belonged to the discoidal hemochorial type, like that of man. The fact that its structure was not villous but lacunal and therefore showed a different circulation seemed even advantageous in this case. The lacunal system facilitates differentiation in morphologically distinct areas, which allow independent functional analysis. The morphological and histochemical studies made during the past decade by Enders (1965), Vollrath (1965), Müller and Fischer (1968), Kaufmann (1969a), Davidoff and Schiebler (1970a, b), King and Enders (1970b, 1971), Davidoff and Gospodinoff (1971), Davidoff (1973), Kaufmann (1974) have led to the discovery of a large amount of new information on this organ, thus rendering it one of the most precisely examined placentae so far. At the same time, however, the hope of finding a model for the human placenta here, has proved vain. It is evident today that findings concerning the guinea-pig placenta are hardly applicable to the human placenta, in particular owing to the completely different blood stream conditions. Even so, the guinea-pig has turned out to be a most important test animal for morphologists, physiologists, biochemists and pharmacologists in experimental work. The great interest of these research groups in exact data on the guinea-pig placenta justifies here a thorough and detailed morphological description. Aspects of more academic interest will have to be dealt with more briefly in favour of those of practical importance. For this reason, its highly complicated embryonic development has only been mentioned so far as is necessary for comprehension of the chapters which follow. In addition, the extensive morphological description of the subplacenta could only be touched upon here, since its practical importance is so far not definitely clear.

While working on this monograph we stated that the representation of the guinea-pig placenta was incomplete in previous publications. Moreover, numerous data demanded reviewing and modernization. The fact that fine structure and volume relations depend decisively on the method of fixation called for systematical comparison of different fixations (Kaiser and Kaufmann, 1976). The picture of the guinea-pig placenta obtained after an optimal perfusion fixation differed so fundamentally from the prevailing conception that this alone demanded a comprehensive representation.

All data and findings in this publication without reference to an author stem from these new studies. As far as the results exceed the framework of this monograph, they will be published in special papers in preparation: ultrastructure of the mature guinea-pig placenta in relation to media and methods of fixation (Kaiser and Kaufmann, 1976); morphometry of the guinea-pig placenta (Stelter and Kaufmann, 1976); development of the placental stem of the guinea-pig (Uhlendorf and Kaufmann, 1976); enzyme histochemistry of the guinea-pig subplacenta (Sax and Kaufmann, 1976); ultrastrucutre of the guinea-pig subplacenta (Wolfer and Kaufmann, 1976).

2. Historical Survey

The first studies on the guinea-pig placenta carried out according to classical histological techniques were of a purely descriptive nature. In 1848 the guinea-pig was chosen by Reichert as an object for his investigations into the implantation of the fertilized oocyte. The main point of his study was the site of the implantation. The first systematic study on the process of implantation was carried out by Bischoff in 1852, at which time he had already handled the development of the placenta. Reichert (1861, 1862) corrected some of the ideas of Bischoff and supplemented them by new data on the development of the amnion.

The investigations which followed, by Bischoff (1866), Hensen (1876, 1883), v. Spee (1883, 1896), Selenka (1884), and Duval (1891, 1892) provided a large quantity of new details, but except for some findings concerning the subplacenta no basically new discovery.

Progress in microscopy from 1900 onwards made more detailed descriptions possible. At first, interest was focused on the early development of the placenta, as is shown by the publications of Spee (1901), Herrmann and Stolper (1904, 1905), Disse (1905), Grosser (1909, 1927), Pytler and Strasser (1925). First functional interpretations could also be made in these early investigations, depending on whether later maturation stages of the placenta were examined (see Blandau, 1949a, b).

In the course of the development of histochemistry, we find the first attempts to define the sites of metabolic processes in the placental tissue. The earliest results were published by Wislocki et al. (1946), who studied the distribution of glycogen, lipids, iron and alkaline and acid phosphatase. At the same time Hard (1946) also studied alkaline phosphatase. More detailed investigations by Vollrath (1965), Christie (1967, 1968), Davidoff, Gospodinoff (1971), and Kaufmann (1974), completed our knowledge of the guinea-pig placenta and allowed a number of functional approaches. The study of the guinea-pig placenta was given new impetus by the introduction of electron microscopy. In particular the studies in this field by Dempsey (1953), Davis et al. (1961a), Enders (1965), Enders and Schlafke (1969), Davidoff and Schiebler (1970a, b), King and Enders (1970b, 1971) should be mentioned. They led to a very precise understanding of the fine structure and ultrastructural development of the different parts of this organ.

A number of biochemical and histo-autoradiographical investigations (Ponse et al., 1954; Jollie, 1964; Stara et al., 1968; Kayden, 1968; Simmer, 1968; Diczfalusy and Mancuso, 1969) brought further information concerning the function of the placenta. The central point of all these considerations was the steroid metabolism.

According to the wish of physiologists who had done intensive work on the guinea-pig placenta since the mid-sixties, numerous investigations were carried out on vascular supply and blood stream direction. The work of Bartels et al. (1967), Fischer (1968), Müller and Fischer (1968), Kaufmann (1969a), Moll et al. (1970), Egund and Carter (1974) should specially be mentioned in this connection. Thanks to these publications, the blood stream conditions of the guinea-pig placenta are probably better known than those of any other placentae.

3. Development of the Guinea-Pig Placenta

3.1. Breeding and Age Determination

The condition for exact morphological, physiological, biochemical and pharmacological studies is the precise determination of the age of gestation. The most common method is the timing of matings.

The beginning of the oestrus of the female guinea-pig can be stated by the desintegration of the vaginal membrane, easily destinguishable outwardly (Spiegel, 1976). The copulation may take place as from the 15th hour of the oestrus onwards (Janiak, 1971). At the end of the oestrus lasting for about 4 days, the vagina is closed by a new epithelial membrane. The whole cycle lasts 16 to 18 days.

However, we consider that planned mating during the oestrus has not proved successful in the case of guinea-pigs owing to numerous reasons:

1) One only obtains a small yield of pregnant guinea-pigs by keeping the animals single and timing matings. Guinea pigs are kept most effectively in small colonies (1 ♂: 4–6 ♀, or a multiple of this) under permanent polygamous conditions. Conception and gestation are then least problematical according to our experience. However, exact determination of the time of conception is then not possible. When keeping pregnant and non-pregnant guinea-pigs together with males, it is important to know that within 24 hours after delivery a postpartum oestrus lasting only 3 1/2 hours occurs (Janiak, 1971). According to our observations 3/4 of the animals kept in colonies conceive during this oestrus. The term of conception is then, cum grano salis, identical with the first day after delivery.

2) The time lapse between mating and conception lies between 6 and 24 hours, or even longer. This means that even with exact knowledge of the time of fertilisation, the calculated age may differ by as much as one day from the actual age of the fetus.

3) It is known that the fetus of mice — similar to that of game (stags, deers and others) — may undergo periods of standstill in the early stages of development (e. g. blastocyst). In examining about 500 guinea-pig pregnancies, we found cases which could only be explained by such a phenomenon. We therefore consider a standstill in development during the early pregnancy of the guinea-pig to be possible although not frequent.

4) Unfavourable conditions for growth may cause local or general retardation of development. When fetuses are implanted close to each other, one is often retarded in its development (if there are three, it is usually the middle one, cp. Ibsen, 1928).

All retardation factors (e. g. unknown lapse of time between mating and implantation, standstill of development in the blastocyst stage, retardation in development owing to deficient nutrition) seem to affect placenta and fetus equally. Not only is the increase in mass of the placenta affected adversely, but its general development also. Dependant on these sources of errors, available data concerning stage and dimensions, in particular in early pregnancy, vary considerably. We therefore consider the use of the conception age not only to be impractical, but also a source of error. We believe the only reliable solution to be the introduction of the "ideal gestation age": i. e., to take as a basis the shortest length of time in which the fetus can reach the respective stages of development, weight and length under ideal conditions. The investigations of Read

9

(1913), Draper (1920) and Ibsen (1928) provide a quantity of material, especially on fetal weight increase. These investigations have shown that the weight increase of the fetus strongly depends on the weight of the mother animal and on the number of fetuses carried. Their use for the determination of the pregnancy age is therefore somewhat doubtful. Due to similar deviations, the weight of the placenta and of the fetal membranes as well as the size of the placenta can also only provide a rough orientation. We believe age determination by means of measuring the crown-rump-length to be the most practicable method, as the fetal length is less dependant on factors such as the size and weight of the mother, or the number of the fetuses, than the fetal weight is.

Certain correlations nevertheless do exist, which may explain differing curves of growth in different studies (Draper, 1920; Kaufmann, 1969a). The breed of the animal may play an important rôle too. The data available for age determination shall be summarized below. All data apply to conventional domesticated guinea-pigs (Cavia porcellus) which are genetically undefined. These bear between one and eight fetuses, on an average 3.3 in our material. In special laboratory breeds we find a substantial increase in the number of young (e. g. in the MRC-breed, the mean is 4.3, with a maximum of 10 to 12 fetuses, Rood and Weir, 1970). Among 249 guinea-pig pregnancies we observed in the years 1970–1973 we found the following distribution of young (Table 1).

Table 1

Number of fetuses	1	2	3	4	5	6	7	8
Frequency	15	57	77	60	25	10	4	1
Percentage	6 %	23 %	31 %	24 %	10 %	4 %	1.5 %	0.5 %

Number of pregnant females examined: 249
Mean number of fetuses: 3.3

Length of pregnancy varies between 59 and 72 days (Rood and Weir, 1970), and in our material it lies between 63 and 66 days for full grown females. Deviations of 2 to 3 days below or above the average are frequent. Shorter pregnancies occur specially with young females. In some studies (Read, 1913; Draper, 1920; Ibsen, 1928), a pregnancy of 68 days is mentioned. These deviations may result from the observation of different breeds.

Since growth is essentially dependant on the weight of the mother, it is advisable to use only full grown animals between 650 and 850 grams. Towards the end of pregnancy they weight between 880 and 1100 grams (with three young). For more details concerning weight increase of the mother during pregnancy see Table 3.

Young females which may first conceive around the end of the second month of age show strong deviations from the norm as to duration of pregnancy, size of the litter, placental and fetal weight. Therefore, they should not be used for scientific purposes.

Measuring the c. r. length of the fetuses presents problems in all stages since the different investigators in this field hardly ever apply exactly the same methods. According to our findings the most constant data are obtained by positioning the animals in all stages of development as follows for measurement: the spinal column is stretched out flat, the hind extremities and the head are bent at right angles. The animals are

Table 2. Age-length relation of guinea-pig fetuses. Our data were obtained by measuring fetuses whose spine was stretched and whose head as well as hind extremities were bent at right angles. The data represent interpolated figures approximated to half a millimeter; they were obtained from an age-length-curve constructed from a population of 240. The measuring conditions of Draper (1920) are not clear. It is likely that he measured fetuses in their naturally curved position

Age in days	Crown-rump-length (Draper, 1920) mm	Crown-rump-length (own measurements) mm	Age in days	Crown-rump-length (Draper, 1920) mm	Crown-rump-length (own measurements) mm
14	2.6	4	39	50.4	48
15	3.2	5	40	53.3	51
16	3.8	6	41	56.3	54
17	4.5	7	42	59.0	56.5
18	5.4	8	43	62.0	59.5
19	6.4	9.5	44	65.0	62
20	7.7	11	45	67.5	65
21	8.9	12.5	46	69.8	67.5
22	10.1	14	47	72.2	70.5
23	11.5	15	48	74.5	73
24	13.0	16.5	49	76.7	76
25	14.8	18	50	78.8	79
26	16.7	19.5	51	80.9	81.5
27	18.8	21	52	82.5	84.5
28	21.0	23	53	84.2	87
29	23.0	25	54	86.0	90
30	25.5	27	55	87.8	93
31	28.1	29	56	89.3	95.5
32	30.0	31	57	90.7	98
33	33.0	33	58	92.2	100.5
34	36.0	35.5	59	93.5	103.5
35	39.0	38	60	94.9	106
36	41.7	40.5	61	96.5	108.5
37	44.3	43	62	97.7	111
38	47.5	45.5	63	99.0	113
			64	100.0	115

then measured from the most prominent point of the head to the stub of the tail. Our results lay above those of Draper (1920) especially in the first and the last third of pregnancy (Table 2). This is due to the fact that the stretching of the spine, in particular in these two stages, leads to a considerable difference compared to the natural position. A full-term, newly born guinea-pig of a 64 day gestational age should have, according to Draper (1920), a length of 100 mm when laid on its side with its spine in the natural curved position. Stretching it, without too strong a traction of the trunk, gave us a considerably higher result of 115 mm. Measurements made earlier than the 14th day of gestation are hardly of any use owing to difficulties in differentiating the fetus from the rest of the embryonic tissue. As weight determination and similar measurements are not applicable here either, one has to resort to structural developmental details in these early stages. This will be dealt with in Table 5.

In the curves on fetal weight development we compared the figures of Draper (1920) with those of Ibsen (1928) (Table 3). In both cases these are the averages from litters of different size. They are therefore only applicable to litters of 3 to 4 animals.

Table 3. Age-weight-relation of mother animal and fetuses during gestation. Weight units in grammes

Age in days	Mean embryonic weight (Draper, 1920)	Mean embryonic weight (Ibsen, 1928)	Mean embryonic weight in the case of extreme number of fetuses (Ibsen, 1928/Kaufmann) 1 fetus–7 fetuses		Weight development of a full grown mother animal carrying 3 young (Read, 1913)
0	0.0	0.0	0.0	– 0.0	751
15	0.0015				
16	0.0022				785
17	0.0050				
18	0.0150				791
19	0.0350				
20	0.0800	0.0581	0.06	– 0.06	785
21	0.140				
22	0.215				795
23	0.295				
24	0.380				851
25	0.470	0.409	0.4	– 0.4	
26	0.570				821
27	0.700				
28	0.880				824
29	1.100				
30	1.500	1.336	1.35	– 1.3	840
31	2.00				
32	2.60				833
33	3.30				
34	4.08				852
35	4.90	4.078	4.5	– 4.0	
36	5.75				867
37	6.65				
38	7.65				883
39	8.80				
40	10.10	9.354	10	– 7	903
41	11.50				
42	13.10				898
43	14.80				
44	16.60				921
45	18.50	20.39	20	– 17	
46	20.50				916
47	22.70				
48	25.00				937
49	27.40				
50	29.90	36.34	36	– 30	972
51	32.50				
52	35.20				979
53	39.00				
54	40.00				980
55	43.90	49.44	52	– 45	
56	47.10				990
57	50.40				
58	53.90				1003
59	57.60				
60	61.60	62.87	70	– 52	971
61	66.00				
62	71.00				990
63	77.00				
64	84.00		90	– 62	988
65		76.52			
66		80.31			988

From our own figures and those of Ibsen (1928) we tried to set up typical weight-spans for litters of 1 to 7 animals in intervals of five days (see Table 3). The numerical data available being very small, the findings can only provide rough guidance.

Data on the size and weight of placentae and fetal membranes can hardly be used for the age determination of the fetuses. For the sake of completeness only they will be quoted here (Table 4). For further details see the notes accompanying the tables.

Table 4. Diameter and weight relations of placenta and fetal membranes. Data on diameter in mm, weight units in grammes

	Weights					Diameters	
Age in days	Placenta and fetal membranes (Draper, 1920)	Placenta and fetal membranes (Ibsen, 1928)	Fetal membranes (Ibsen, 1928)	Placenta (Ibsen, 1928)	Placenta in the case of 1–6 young (Ibsen, 1928/ Kaufmann)	Main placenta (Uhlendorf, Kaufmann, 1976)	placental stem (Uhlendorf Kaufmann, 1976)
15	0.6						
20	0.8			0.123		8.5	7.4
25	1.1	0.433	0.070	0.363	0.6 – 0.3	13	8.5
30	2.2	0.933	0.208	0.725	1.0 – 0.5	17.5	9.5
35	3.5	1.440	0.343	1.097	1.7 – 1.0	20.5	10.0
40	4.5	3.195	0.745	2.45	2.6 – 1.7	22	10.4
45	5.3	4.379	1.02	3.359	4.0 – 2.5	23	10.2
50	5.7	5.33	1.38	3.95	5.3 – 3.5	24	9.2
55	6.0	6.39	1.56	4.73	6.0 – 4.0	24	8.0
60	6.3	6.704	1.61	5.094	6.7 – 4.0	24	8.0
64	6.5	7.14	2.19	4.95	7.3 – 3.3	25	8.0

For control of the gestational age and especially for the choice of animals in stages suitable for the respective investigations, the following factors are indispensible: verification of the pregnancy of the mother animal, determination of the number of fetuses and of their developmental stage. Our experience proved the following examination technique to be successful (Fig. 1): for the examination of the right uterus horn the mother animal is caught by the back of the neck with the thumb, index and middle finger of the left hand. The other two fingers hold the thorax, thus fixing it against the ball of the thumb. The front extremities are held in this way between the middle and the ring finger. The animal is then turned onto its back and the lower part of the body is put on a firm surface. The right hand now grasps the abdomen from the right side, the thumb pressing the spine dorsally and the four finger tips ventrally. The fingers are spread out and drawn together in a continuous movement while the hand is being drawn laterally across the abdomen. In this manner the uterus should be felt between thumb and finger tips sooner or later, and swellings which indicate a pregnancy can be detected. For the examination of the left uterus horn, the animal is held with

13

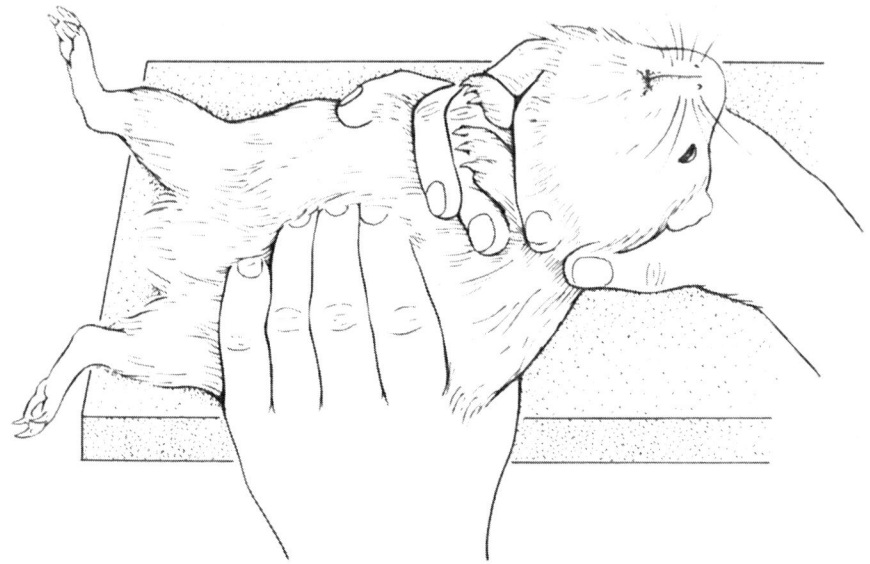

Fig. 1. Schematic representation of the examination of the right uterus horn of a pregnant guinea-pig. The mother animal is fixed with the right hand. Index finger, middle finger and ring finger of the left hand palpate the uterus horn pressing it through the abdominal wall against the left thumb. The pelvis of the mother animal is propped against a firm surface. The examination of the right uterus horn is carried out in similar fashion

the right hand and palpated with the left. An inexperienced examiner may palpate the kidneys in the upper abdomen by mistake.

According to size and quality of the palpational findings on the right and on the left side of the abdomen, the pregnancy can be classified in 6 stages without difficulty (Fig. 2): in stage 1 (up to the 15th day), there are peasized swellings (up to 6 mm diameter). The earliest pregnancy stages we were able to verify in this way lay around the 10th day. In stage 2 (15th to 25th day of gestation) we find firm ballshaped swellings of the size of hazelnuts in the region of the uterus horns (7 to 15 mm diameter). In stage 3, the palpational findings are firm, elastic, slightly oval bodies nearly of chestnut size (15 to 30 mm diameter). In the stages 1, 2 and 3 parts of the body cannot yet be differentiated by palpation.

Typical of the following 3 stages, i. e. the second half of pregnancy, is that, instead of firm and elastic bodies, one can now palpate parts of the fetus, especially the head and pelvic regions, owing to the relative reduction of amnion fluid. This may be another source of error, as the head may be taken for a single embryo of an earlier stage. The hard and knobby consistency, however, differs from the firm and elastic consistency of the earlier stages, which is a help in diagnosis.

Stages 4 to 6 can be differentiated as follows: in stage 4, we find cylinder-shaped bodies of 3.5 to 5 cm length with two more solid parts (head and pelvis) and a softer constricted middle part. In stage 5, the fetuses have a palpable length of 5 to 7 cm, and head, thorax, and pelvis are easy to distinguish. In stage 6, the fetuses have a length of 7 to 10 cm as judged by palpation, and a hard, knobby head of approximately 2.5 cm length.

14

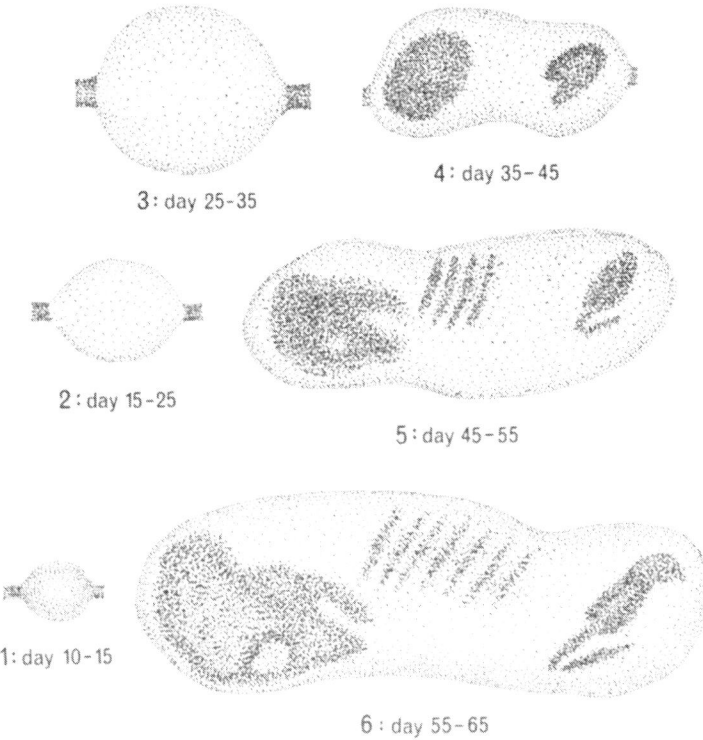

3 : day 25-35

4 : day 35-45

2 : day 15-25

5 : day 45-55

1 : day 10-15

6 : day 55-65

Fig. 2. Palpational findings in pregnancy examination in original size. Earlier than the 10th day the uterus horns show no swellings. In pregnancy stages 1 to 3 (10th to 35th day) one feels firm-elastic bodies (dotted) in the uterus horn. In stages 4 to 6 (35th to 65th day), hard fetal body parts can be differentiated in terms of palpation in the now flaccid amniotic cavity

This method of palpation allows a sufficiently safe estimation of gestational age with an accuracy factor of ± 5 days and, with practice, of ± 3 days. Possible errors are that the kidneys or bladder may be mistaken for a fetus, or that head and pelvis are taken for fetuses of earlier stages. Any firm and elastic body of hazelnut size near the median line in the lower abdomen should always be suspected to be a full bladder. A slight pressure which causes emptying of the bladder is decisive in diagnosing the difference. One ought to be careful when palpating the upper abdomen as too strong a pressure on the kidneys or too strong a traction on the renal vessels may result in an intraabdominal bleeding. The approach of term is easiest to state by palpating the symphysis (Spiegel, 1976). The pubic bones start to part 3 to 4 days before term and stand 1 to 2 cm apart at term.

3.2. Early Development and Formal Genesis

According to Pytler and Strasser (1925), the oocyte is fertilized 6 to 12 hours after the coitus or even later. The fertilized oocyte reaches the implantation site in the uterus in the stage of a morula or early blastocyst between the 4th and the 5th day (Bischoff,

1852; Hensen, 1876, 1883; Creighton, 1878; Herrmann and Stolper, 1904, 1905; Disse, 1905; Amoroso, 1952). At this point the conceptus is still covered by the zona pellucida. Enders and Schlafke (1969) speak of a non-implantation-stage which passes into the appositional stage after the attachment of the blastocyst to the uterine wall. According to v. Spee (1896, 1901) and Blandau (1949a) projections of the trophoblast may break through the zona pellucida and penetrate into the decidua at this stage already. The zona pellucida vanishes only after the 6th and 7th day. Streamers of trophoblast now invade the decidua in greater quantities. Enders and Schlafke (1969) call this the adhesional stage of the implantation. During implantation, the blastocyst acquires a cylindrical shape with the axis parallel to the longitudinal axis of the uterus.

Implantation normally takes place antimesometrially. At the end of the 7th day the blastocyst has penetrated so deeply into the decidua that the mucosa has closed again over it (Fig. 3a). This causes a big bulge in the endometrial epithelium which is referred to as the nidus formation. Within the blastocyst, two different zones can be discerned (Grosser 1909, 1927; Amoroso, 1952): one is the single-layered blastocystic wall which undergoes a retrogressive metamorphosis during the following days. An accumulation of cells is attached to its inner side mesometrially, i. e. on its anti-implantation pole, from which placenta and embryo later on originate. Within this cell accumulation, the differentiation of the cell layer facing the blastocystic cavity into the endoderm starts at the beginning of the 8th day. Some authors have later dates, for example Amoroso (1952) and Starck (1975), name the 11th day. With the growth of the blastocyst, the blastocystic wall starts to recede in spots. The ectoderm lying between the endoderm and the blastocystic wall proliferates to form the so-called egg cylinder which proceeds, still completely covered by the endoderm, into the blastocystic cavity (Fig. 3b). Analogous to other rodents, this superficial endoderm layer is termed visceral yolk sac wall. Further development in the case of most mammals (with the exception of rodents and lagomorphs) takes one of two forms. Either a cavity is formed in this newly built endoderm layer and extends until the developing endoderm layer finally fills the blastocystic cavity as yolk sac, or the surface of the endoderm layer extends from the embryoblast to the trophoblastic wall, lining the whole blastocystic cavity until it finally closes as yolk sac. In both cases, the part of the yolk sac epithelium which is attached to the embryoblast is referred to as visceral yolk sac epithelium and the part lining the inside of the blastocystic wall as parietal yolk sac epithelium.

In the case of guinea-pigs (and other rodents as well as lagomorphs), the yolk sac epithelium no longer extends from the embryoblast on to the blastocystic wall as the latter largely recedes with the exception of the ectoplacental cone region. Thus only the visceral wall of the yolk sac is formed. A yolk sac cavity exists in no stage of the development.

Within the ectoderm covered by the yolk sac one can differentiate two sections on the 8th day: 1. the träger or ectoplacental cone of Duval (1891) — the part facing the uterine cavity largely consisting of syncytium and lined by a thin cytotrophoblastic layer on the surface facing the blastocystic cavity — and 2. the embryoblast which lies still nearer to the blastocystic cavity than this cytotrophoblast (Fig. 3b). Further growth of the egg cylinder causes narrowing of the lumen of the uterus. Antimesometrial of the blastocyst a cleft now occurs which does not yet have a lumen at this stage.

On the 9th day, the uterine lumen has decreased to a needlesized channel, whereas the new uterine lumen which has just formed antimesometrially of the embryo increases

Table 5. Derivatives of the trophoblast and the embryoblast

Day							
6	decidua		blastocyst				
7			trophoblast (syncyto- and cytotrophoblast)		embryoblast		
8			ectoplacental cone (Träger)		embryoblast	endoderm (visceral yolk sac epithelium)	
9		basal ectoplacental cone	egg cylinder	extraembryonic mesoblast	germ-amnion-sac		
10	syncytial sprouts	trophospongium	central excavation	allantois	connective stalk	amnion mesoderm	amnion epithelium	embryonal mesoderm	embryonal ectoderm	embryonal endoderm	yolk sac mesoderm
11	junctional zone						
12				fetal vascularization		amnion	embryo	yolk sac placenta
13							
14							
15				roof of the central excavation			
16	decidua	junctional interlobium labyrinth		subplacenta	umbilical cord	yolk sac stalk	amnion	embryo	yolk sac placenta
↓ 64		main placenta					

17

(Fig. 3c). From this surface, single glands grow into the decidua. In the egg-cylinder, the ectoplacental cone and the embryoblast separate and grow away from each other. The narrow space between them – the proexocoel (Pytler and Strasser, 1925) – is filled up with a primitive connective tissue, the extraembryonic mesoblast. This stems largely from the cytotrophoblast layer covering the ectoplacental cone. The embryoblast, or more precisely, the germ-amnion-ectoderm, shows in its center a focus of disintegration and acquires an increasingly vesicular structure. The extraembryonic mesoblast between the embryoblast and the visceral yolk sac wall continues growing until both are completely separated (Fig. 3c).

Also within the ectoplacental cone a cavity is formed by tissue disintegration – the ectoplacental cavity – which however does not appear to have great significance for further development.

After the blastocystic wall has involuted completely, the visceral layer of the yolk sac grows around the ectoplacental cone thus loosening it from the decidua. This is here designated ectoplacental endoderm and forms the outmost layer of the conceptus as from this stage. The phenomenon that now the ectoderm is enveloped by the endoderm is known as the inversion of the germ layers. In this way, the visceral yolk sac epithelium assumes the rôle played by the chorion laeve in many other mammals whose yolk sac undergoes involution while the blastocystic wall remains.

Clefts occur now in the extraembryonic mesoderm (Fig. 3c) and soon join together to form a bigger cavity (Fig. 3d): the extraembryonic coelom or the exocoel. At the base of the ectoplacental cone projections of syncytium break through the endoderm and penetrate into the endometrial decidua. At this place the "Durchdringungszone", or the junctional zone develops, characterized by fetal and maternal necrosis of tissue (Bischoff, 1866; Ercolani, 1879; Laulanié, 1886; Grosser, 1909; Davies et al., 1961a).

Between the 9th day and 10th day (Fig. 3d), the original uterine lumen has disappeared completely, the newly formed antimesometrial uterine lumen spreads continually to surround the germ sac and in this way separates a parietal from a capsular layer of decidua. A narrow space (the decidual cavity) remains free between the outmost fetal membrane (the yolk sac) and the decidua in most places, with the exception of the mesometrial junctional zone. The extraembryonic coelom extends further and presses the embryonic mesoblast as a thin layer of connective tissue against the endodermal wall. Thereby the coelom grows on the one hand around the germ-amnion-sac, on the other it begins to proceed into the depth of the ectoplacental cone in the shape of a wedge. The ectoplacental cavity is deformed to form a cleft only. An extension of the mesoderm grows out of the marginal zone between germ-amnion-sac and yolk sac towards the ectoplacental cone (Fig. 3d). This extension is the forerunner of the allantois. At the embryonic pole of the germ sac, the wall of the yolk sac forms a groove towards the germ-amnion-sac which then will become the intestinal groove.

On the 10th day (Fig. 3e) the intestinal groove becomes deeper and continues growing in the direction of the germ-amnion-sac, moving it into the amniotic cavity and displacing the extraembryonic mesoblast. At that point the allantois is firmly attached to the ectoplacental cone. The germ-amnion-sac which is now connected with the yolk sac starts moving relative to the wall of the amniotic cavity and towards the placenta owing to uneven growth of the yolk sac epithelium. Simultaneously, syncytium sprouts spread from the ectoplacental cone into the junctional zone. Here they ramify like roots. The extraembryonic mesoblast growing wedge-shaped into the ectoplacental cone (appearance of the central excavation) restricts the ectoplacental cavity which finally disappears (Fig. 3e).

In the course of the 11th day the allantois fuses with the ectoplacental cone (Fig. 3f). Allantois mesoderm with a high number of cells invades the central excavation of the placenta which thereupon spreads and deepens. In the previous days, focusses of desintegration have appeared in the syncytium of the placenta transforming it into a spongelike area (trophospongium). Owing to an exceptionally intense proliferation of the basal processes of trophoblast in the junctional zone, maternal vessels in the endometrium are eroded. Their blood flows into the channel network of the trophospongium (Duval, 1892). According to Pytler and Strasser (1925) this happens on the 12th day whereas Franke (1969) placed it on the 9th day.

On the 12th day the rims of the intestinal groove fuse to form the intestinal tube, and the intestinal endoderm detaches from the yolk sac epithelium (Fig. 3f). The invagination of the yolk sac epithelium is gradually retracted in the direction of the surface of the conceptus. However, a mesoderm bridge (the so-called body stalk or connective stalk) between the wall of the embryo and the germ-amnion-sac remains. On the same day, within the allantois mesoderm and also in the mesoderm of the connective stalk, fetal blood vessels can be detected, which grow via the allantois to the placenta or via the body stalk to the yolk sac wall respectively.

On the 13th day, the trophoblast of the placenta-forming tissue has differentiated mostly to syncytium. Cytotrophoblast is only to be found on the fetal surface of the placenta, on the outer surface of the placenta beneath the ectoplacental ectoderm which at this stage already no longer forms a continual layer, and at the base of the central excavation. Starting from the central excavation, mesoderm invades the syncytium radially.

On the 14th day, the base of the central excavation extends, acquiring a fungoid shape (Fig. 3g). The connection between the central excavation and the exocoel is constricted by growths of the trophoblast. The layer of cytotrophoblast lining the base of the central excavation is now designated the "roof" of the central excavation (Duval, 1892). The mesoderm of the roof of the central excavation invades the cytotrophoblast in the form of lamellae and also the syncytium from which it remains separated by a layer of cytotrophoblast. The roof of the central excavation is the forerunner of the subplacenta. The trophoblast lying laterally of the central excavation will form the main placenta (Kaufmann, 1969a).

On the 15th day, the central excavation becomes more constricted on its fetal side owing to the massive growth of trophoblast (Fig. 3g). However, it will persist until the end of the pregnancy as a cordlike, thin connection (central mesenchymal axis) between allantois, the later umbilical cord and the roof of the central excavation.

At this point the germ-amnion-sac has completed its rotation of 180° and accordingly the originally antimesometrial located umbilical region of the embryo now points towards the placenta. Due to mass moving in the sphere of the allantois and in the mesoderm of the connective stalk, the latter looses its direct contact with the germ-amnion-sac and is now connected with the allantois, the later umbilical cord. At this stage, the placenta-forming tissue undergoes an overproportional growth and, simultaneously, a change in shape, from that of a broad-based wedge to that of an ellipsoid. Owing to this growth, the originally roughly spherical yolk sac suffers a strong swelling at its placental pole. The fusion site of the yolk sac epithelium and the placenta is thereby displaced from the rim of the placenta further in the direction of the fetal surface of the

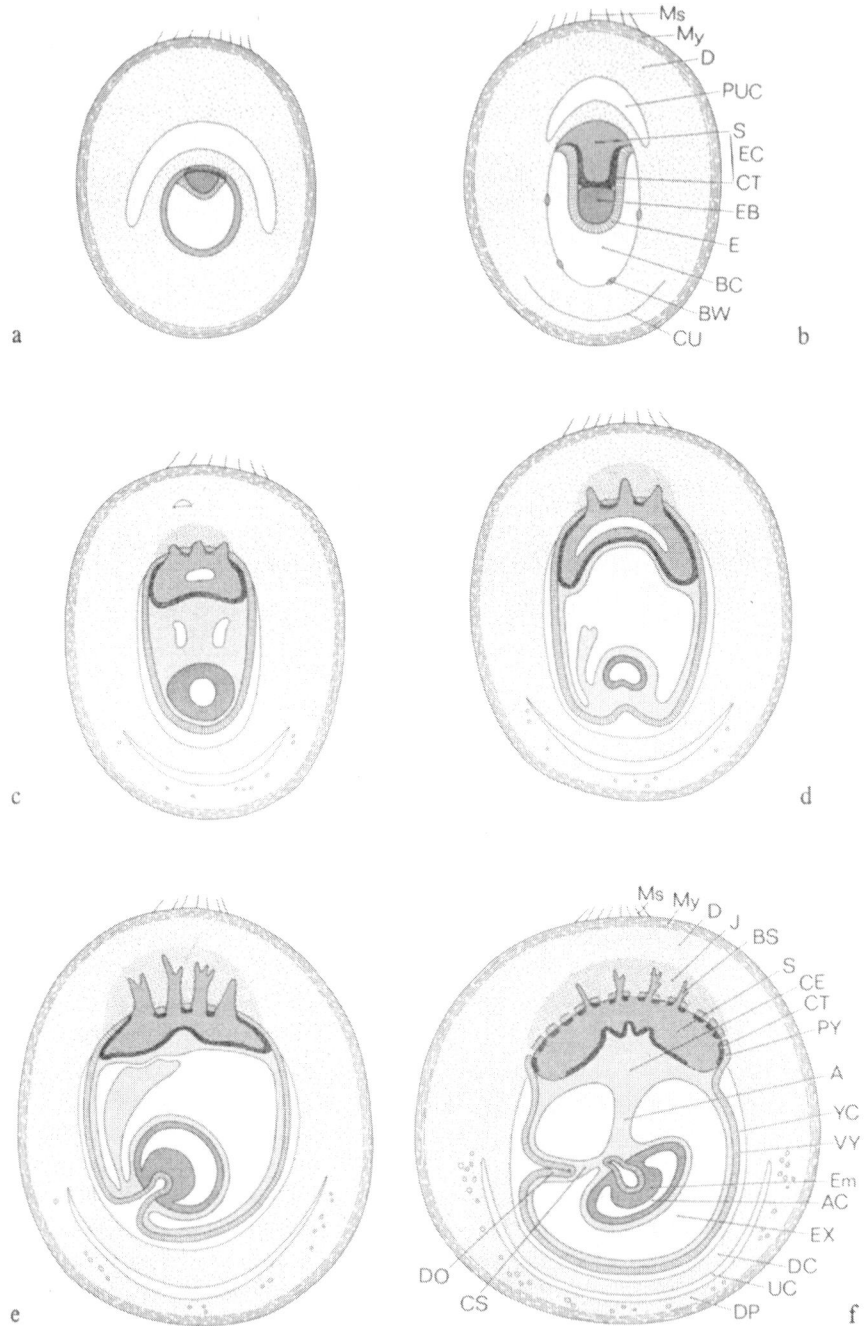

Fig. 3a–i. Strongly schematic representation of the development of placenta, fetal membranes and embryo of the guinea-pig, based on transversal sections through the uterus horn at the site of the middle of the placenta. (a) 7th to 8th day; (b) 8th day; (c) 9th day; (d) 9th to 10th day; (e) 10th day; (f) 12th day; (g) 15th day; (h) 20th to 32nd day; (i) 63rd day. In the schemes a–h, a true-to-scale representation was neglected for better schematic clarity. Only the representation in scheme i (guinea-pig pregnancy at term) is true-to-scale (X 2,75).

A = allantois; AC = amniotic cavity; Am = amnion; BC = blastocystic cavity; BS = basal syncytial sprouts; BW = blastocystic wall; CE = central excavation; CS = connective stalk; CU = cleft of the

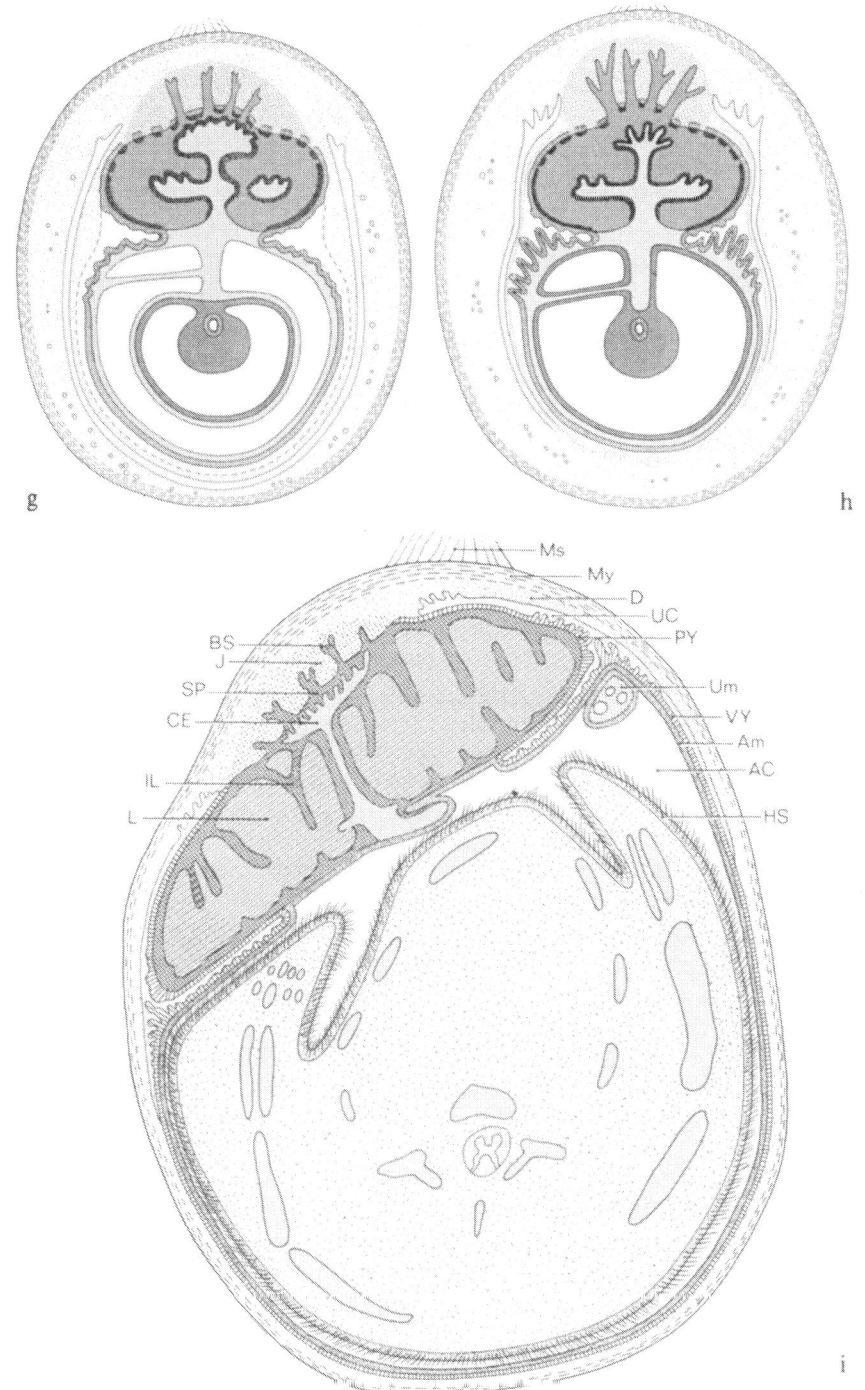

secondary uterine lumen; *D* = decidua; *DC* = decidua capsularis; *DO* = ductus omphaloentericus; *DP* = decidua parietalis; *E* = endoderm; *EB* = embryoblast; *EC* = ectoplacental cone; *Em* = embryo; *EX* = exocoel; *HS* = haired skin of the fetus; *IL* = interlobium; *J* = junctional zone; *L* = labyrinth; *Ms* = mesometrium; *My* = myometrium; *PUC* = primary uterine cavity; *PY* = parietal yolk sac; *S* = syncytotrophoblast; *SP* = subplacenta; *UC* = uterine cavity; *Um* = umbilical cord; *VY* = visceral yolk sac; *YC* = yolk sac cavity placenta. A considerable surface extension of the yolk sac epithelium in close vicinity to its attachment site on the placenta leads to an intensive formation of villi. The yolk sac epithelium adhering to the placenta (the ectoplacental endoderm) also starts forming villi to a lesser extent.

On this and the following day, the newly built secondary uterine cavity expands continously towards the mesometrium, encompassing the embryo, until it finally reaches the region of the junctional zone. With the growth of the embryo the amniotic cavity also increases at the cost of the extraembryonic coelom (exocoel).

Around the 17th day the amniotic cavity has expanded to the extent that the two-layered amnion — consisting of an inner ectodermal and an outer mesodermal layer — meets large areas of the mesodermal lamella which lines the yolk sac epithelium inwardly. The allantois stem and the connective stalk are enveloped by this two-layered amnion lamella as well. The extraembryonic coelom disappears leaving behind a cleft only. The villous foldings of the yolk sac epithelium close to the attachment site of the placenta grow, thus filling the extra space in the original decidual cavity which resulted from the constriction of the germ-amnion-sac (Fig. 3h). During the following period there occur only insignificant changes of the fetal membranes: amnion mesoderm and yolk sac mesoderm fuse locally, the separating mesothelium disappearing. The decidua capsularis outside the yolk sac epithelium also thins more and more and finally disintegrates completely in the second half of the pregnancy so that the yolk sac epithelium comes into direct contact with the decidua parietalis (Fig. 3i). The secondary uterine lumen expands at the level of the subplacenta under the main placenta. Thus the placental contact area with the decidua diminuishes steadily until its connection with the uterine wall has reduced to a stem only. At the end of gestation, the diameter of this stem measures only a third of that of the main placenta (Fig. 4).

3.3. Morphogenesis

On the 14th day of gestation, the guinea-pig placenta has acquired the shape of a broad-based cone which is embedded in the decidua with its tip pointing towards the myometrium. The inside of this cone (the central excavation) is filled with fetal mesoderm and is in contact with the allantois (Fig. 3f). The cone mantle consists of a network of trabeculae and plates of syncytotrophoblast between which the maternal blood is flowing. The tip of the cone is formed by a thick, folded layer of cytotrophoblast (towards the fetal side) and by syncytium (towards the decidua). A few longish and branched syncytial projections invade the decidua from the tip of this cone. In sections they sometimes seem to lose continuity with the placenta which justifies accordingly their designation as syncytial giant cells. Serial sections, however, generally reveal the existence of continuity with the basal syncytial sprouts. These syncytial sprouts perforate a practically complete layer of ectoplacental endoderm at the base of the placenta.

All that time, cellular trophoblast is also to be encountered in the remaining areas of the placenta — mostly superficially below the yolk sac epithelium cover as well as in the border area between syncytium and fetal mesoderm of the central excavation. Originating from the central excavation, mesenchymal cells invade the syncytium in which they hollow out the syncytial trabeculae paralleling the maternal blood lacunae.

Davidoff and Schiebler (1970b) already observed first penetration of fetal vessels in the syncytium at this stage of development.

On the 15th and 16th day, the top of the central excavation widens, and fetal mesoderm grows from this point between the lamellae of the bordering folded cytotrophoblast. This is the beginning of the typical subplacenta (= roof of the central excavation (Duval, 1892)). The part of the placenta remaining after differentiation of the subplacenta and lying laterally of the central excavation is termed main placenta. Owing to intensive growth of the syncytotrophoblast of the main placenta, the part of the central excavation in the vicinity of the fetus narrows (Fig. 3g). Only a cord-like connection remains between the area of the later insertion of the umbilical cord and the roof of the central excavation. On the 16th day, the fetal mesenchyme increasingly invades the syncytium radially from this central mesenchymal axis (Fig. 3g, h). This process is particularly pronounced in the third of the placenta distal from the uterus, so that a wheel-like plate of syncytium interspersed by connective tissue is formed here (labyrinth plate). In the syncytium, fetal capillaries containing nucleated erythrocytes are first found by means of the light microscope (Kaufmann, 1969a) at this stage (16th to 17th day). The cytotrophoblast can now only be traced peripherally under the yolk sac epithelium. It is separated from the bordering syncytium by a compact homogenous and strongly basophilic streak of syncytium which remains up to the 20th day. The basal part of the cytotrophoblast proximate to the decidua begins transforming into giant cells.

As from the 17th day, conical bodies grow from the newly built labyrinth-plate in the top third of the placenta in the direction of the uterus. From these cones, the mesenchyme spreads radially into the syncytial spongium. On the 21st day, several are already fully developed. Shorter conical areas growing off the labyrinth-plate and pointing towards the embryo are vascularized fetally as well. The cytotrophoblast lying peripherally between the surface of the placenta and the ectoplacental endoderm has been almost completely transformed into a multi-layered stratum of Duval's giant cells. These begin to secrete a homogenous eosinophilic mass into their surroundings, preferably towards the yolk sac. The Reichert membrane develops from this mass until the 23rd day. At that time, the number of giant cells diminishes markedly. On the 32nd day only a fragmentary layer of giant cells is left, and the formerly highly prismatic visceral yolk sac epithelium starts involution, at first at the pole of the amniotic cavity opposite the placenta. After disappearance of the major part of the villi and involution of most of the fetal capillaries which had come via the yolk sac stem (the former connective stalk), the epithelium thins to a single-layered stratum of plated epithelium. This process continues until term: at birth only the yolk sac cover of the main placenta and a bordering, approximately 1 cm broad stripe of the free yolk sac still consist of highly prismatic, folded, villous yolk sac epithelium (see Dempsey, 1953; Petry and Kühnel, 1963) (Fig. 21a–e).

With the invasion of fetal mesenchyme (from the 14th till the 16th day (Fig. 12), striking changes occur in the syncytotrophoblast. The originally very large nuclei start dividing very rapidly near the mesenchyme and thus lose volume. The syncytial trabeculae also divide, producing a close-meshed network with numerous small nuclei. These align in ultimate vicinity to the mesenchyme parallel to the direction of growth of the mesenchyme. It is noteworthy that this new abundance of nuclei does not extend to the recently vasculated region. Many syncytial nuclei seem to perish when penetrating mesenchyme and fetal capillaries (see electron microscope findings, chapter

6.2.1). The very loose fetal mesoderm in whose periphery at first no capillaries can be traced light-microscopically moves deeper into the syncytium, cleaving the syncytial trabeculae and plates. Evidently, maternal blood spaces are not opened thereby. According to our light-microscopical findings, the capillarization of areas opened by fetal mesenchyme follows very rapidly. It is nearly accomplished by the 23rd day so that capillaries then border directly on the unopened regions of syncytium. Further expansion no longer takes its origin from a uniform mesenchymal front but from individual fetal capillaries some of which penetrate deeply into the syncytium not yet fetally vascularized (interlobar and marginal syncytium). Individual fetal capillaries of this type are frequently to be found in the interlobar and marginal syncytium up to the 45th day, thus demonstrating a continual spreading of the labyrinth at the cost of uncapillarized trophoblast. Even in mature placentae fetal capillaries may occasionally be detected in these zones. This finding is in accordance with the fact that the relative proportions of fetally vascularized to unvascularized syncytium continually alter towards term in favour of the former (cp. Table 7).

As is further evident from Table 7, the typical tissue of the lobar centre already forms on the 16th day. It consists of very large maternal blood lacunae in thick syncytial trabeculae with honeycomb-textured cytoplasm which is mostly slightly basophilic. The development of this region is particularly intensive during the first days. Already on the 20th day of gestation it constitutes approximately 11% of the main placenta's volume and thus a considerable part of the labyrinth. In contrast to the labyrinth as a whole which occupies a continually increasing portion of the placental volume, the share of the lobar centre in the placenta remains practically constant from the 20th day to term.

In the very last days of gestation – usually together with the widening of the symphysis – white infarctions occur to a varying degree in the main placenta. They prevail in the basal parts of the placenta laterally of the stem. They occur more seldom in the rest of the placenta, either disseminated in pin-point size or as a wedge-shaped infarction. They hardly ever are to be found before the 60th day of gestation. According to our own earlier investigations (Kaufmann, 1975b), the formation of such a fibrinoid infarction takes approximately 10 days. The cause for this phenomenon should therefore be sought between the 50th and 55th day of gestation.

4. Rough Structure of the Mature Placenta

As from the middle of gestation, the guinea-pig placenta with its three component parts: main placenta, placental stem with subplacenta and the yolk sac placenta, are largely developed. Until term, they only alter in terms of size and relative volume proportions.

At term, the main placenta has the form of a thick, slightly oval disc (Fig. 4). From the insertion of the yolk sac on the fetal surface of the placenta up to the endometrium, the placenta has a three-layered mantle, consisting of the yolk sac epithelium on the outside (ectoplacental endoderm), the Reichert membrane and the Duval giant cell layer on the inside. On the fetal surface within the yolk sac insertion, the main placenta

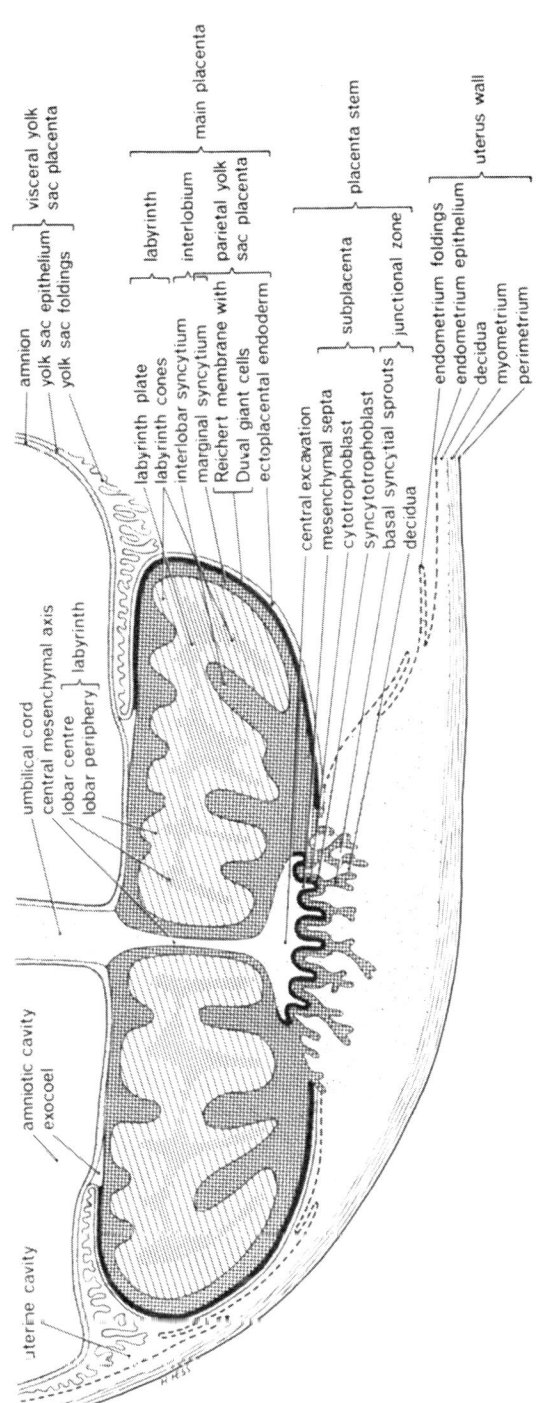

Fig. 4. Schematic, vertical section through a nearly mature guinea-pig placenta (not true-to-scale) with bordering parts of uterus wall and fetal membranes. Whereas fetal membranes and uterus wall are left in their natural vicinity to the placenta in the left half of the picture they were separated on the right in a way common after opening of the uterus

is covered only by very thin single-layered amniotic epithelium and some connective tissue.

The remaining tissue of the main placenta can be subdivided into non-capillarized syncytium (peripheral situated marginal syncytium, interlobar syncytium between the lobes) and capillarized syncytium (the labyrinth). The structure of the labyrinth, explicable in terms of development, is as follows: in the upper third (fetal end) of the main placenta the capillarized syncytium forms a plate (labyrinth plate) paralleling the surface of the uterus and occupying the major part of the space in question (Fig. 4,9a). This plate is run through radially by streamers of non-capillarized trophoblast (Fig. 9b). In its centre it is perforated by an axis of vessels and connective tissue (central mesenchymal axis), remnants of the central excavation (Fig. 4). Cones of labyrinthine tissue, very short ones towards the fetus (Fig. 4, 9a, 11a) and longer, thick ones towards the decidua (Fig. 9a, c, 11b), run from the plate parallel to this axis. They may branch, but may also fuse, the intermediate interlobar syncytium disappearing. In a cross-section (or horizontal section of the placenta), the cones reveal the typical lobular structure of the guinea-pig placenta (Fig. 9c). However, the lobes or cross-sections of the cones are roughly round only in the centre of the main placenta, whereas they become elliptical towards the surface. In the centre of the lobes a loose, basophilic tissue is conspicuous (Fig. 10, 13e, f). It is composed of considerably widened maternal blood lacunae, whose syncytial walls are thicker here than in the periphery of the lobe, and fetal vessels which are widened as well. This tissue is termed lobar centre (Kaufmann, 1969a). It cannot be sharply distinguished from the adjacent zone of the lobar centre, the so-called lobar periphery, starting in the outer third of each lobe (Fig. 10, 13c, d). The region between lobar periphery and interlobar or marginal syncytium is designated transitional zone (Fig. 10). Since interlobar and marginal syncytium show no differences, either in terms of ontogenesis, light or electron microscopy, we refer to them without distinction as the interlobe.

The discoidal main placenta is connected with the uterine wall by the placental stem. Around the end of gestation, this stem only measures approximately 8mm, which is only just about one third of the diameter of the main placenta (25mm). The stem consists mainly of the subplacenta (see chapter 6.4), which is surrounded by a narrow rim of junctional zone tissue. Big maternal arteries and veins run through the latter into the main placenta. As from the transition from the placental stem to the main placenta, the main placenta is covered by yolk sac towards the fetus. The yolk sac covers all the lateral parts of the main placenta. It detaches itself only from the placental surface pointing towards the fetus at a distance of 1/2 to 1 cm from the insertion of the umbilical cord to cover the amniotic cavity outwardly as free yolk sac (see chapter 6.5).

5. Vascular Supply of the Mature Placenta

As the vessels leading to the placenta are almost of greater importance for the scientist working experimentally on guinea-pig placenta than the intraplacental vessels (Dancis, 1964; Dancis and Money, 1960; Künzel and Moll, 1972; Money and Dancis 1960), the former will be described first.

5.1. Maternal Arterial Blood Supply

The two uterus horns are supplied with blood by an arterial loop approximately 8 to 10 cm long which extends between the ovarian and the uterine arteries (Fig. 5a). It is named the arcade artery and runs through the mesometrium. Whereas the main supply of this arcade artery is provided for by the ovarian artery in non-pregnant animals and also those in early pregnancy, the diameter of the uterine artery increases with gestation age and carries about 60 % of the blood supply of the uterus towards term, according to Moll and Künzel (1971). As shown by our vessel casts (Fig. 5a) and the angiografical studies of Egund and Carter (1974), there are no significant anastomoses between the arcade arteries of both sides, as described by Schniewind and Asshauer (1962), Thomsen et al. (1966) as well as by Fischer (1968). These authors incorrectly stated a connection of the two ovarian arteries by the arcade arteries. Approximately one dozen radial arteries measuring 3 to 4 cm in length appear on either side of these arcade arteries and enter the uterine wall via the mesometrium (Fig. 5a). Each placenta is supplied by 1 to 3 of such radial arteries. Those running through the uterine wall into the placenta are characterized by the fact that they have a diameter of only 0.5 mm or even less at the point where they branch off the arcade artery. Shortly before the uterus wall, however, they become club-shaped and reach a diameter of 1 to 1.5 mm (Fig. 5a). According to Egund and Carter (1974), the purpose of this arterial dilatation is the reduction of blood speed and blood flow before placental perfusion. At the site of the placenta, the radial arteries lying tangentially in the uterus wall turn at right angles and run in numerous windings — as intramural or spiral arteries — through the uterus wall into the stem of the placenta (Fig. 5b). On the way the vessel itself may bifurcate. Smaller vessels branch off to supply the uterus wall.

Physiological studies of Leichtweiß and Schröder (1976) revealed that 80–92 % of the blood volume circulating through the radial arteries running to a placenta, does actually flow through the placenta. The remaining 8–20 % flow via the above mentioned small intramural branches into the myometrium. In artificial maternal perfusion of the placenta we therefore have here a shunt of importance for all calculations.

In the placental stem, one usually finds 3 to 4 large caliber arteries arranged in a circle around the subplacenta. Very fine vessels branch off at this point to ensure the arterial supply of the subplacenta (Fig. 5b). Having passed the junctional zone situated laterally to the subplacenta, they reach the main placenta. Contrary to our earlier descriptions (Kaufmann, 1969a), they do not ascend in the interlobium, but enter basally into some central labyrinth lobes. They break loose from their arterial wall at this transitional point so that the maternal arterial blood now flows in main lacunae with trophoblastic walls. Owing to further ramification 4 to 6 ascending lacunae are formed. They give off capillary lacunae radially into the surrounding labyrinth lobes thus ensuring the arterial supply of these lobes. One spiral artery usually supplies only one sector of the main placenta (Fig. 5b) since the 3 to 4 arterial main lacunae with their 4 to 6 branches show but very small anastomosing branches; only very small anastomoses occur in the area between subplacenta and main placenta and none at all in the placenta stem between the spiral arteries. Accordingly in experiments, the placenta as a whole cannot be perfused through a single radial artery unless all spiral arteries branch from this one radial artery in question.

The 4 to 6 main lacunae ascending in the central labyrinth lobes traverse approximately two thirds of the thickness of the placenta. Then they turn, taking a horizon-

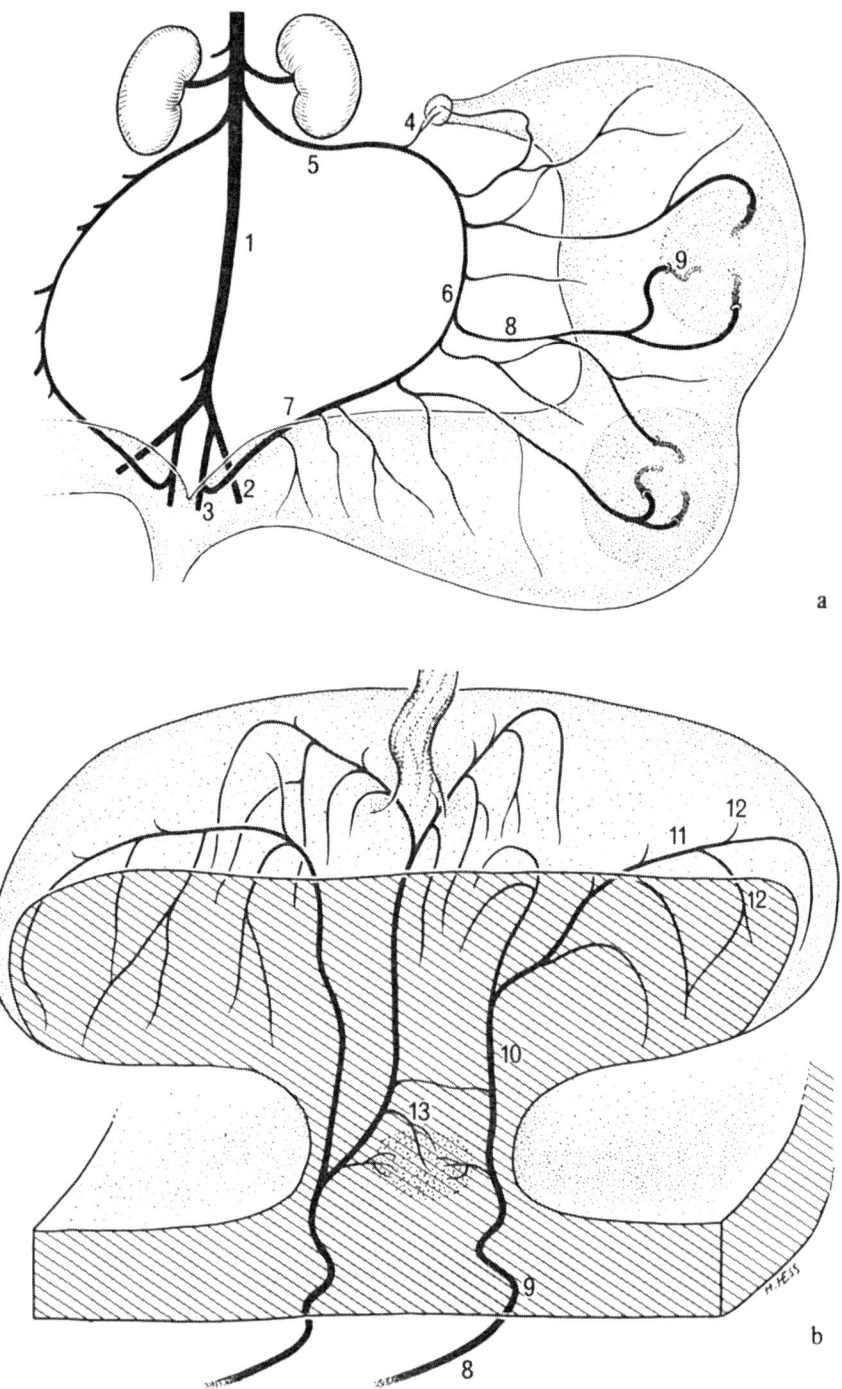

Fig. 5a and b. Schematic, largely true-to-scale representation of the maternal arteries in the second half of gestation. (a) supplying maternal arterial system, x1. (b) intramural arteries and intraplacental arterial lacunal system, X 5.

1: abdominal aorta; *2*: a. iliaca externa; *3*: a. iliaca interna; *4*: ramus ovaricus a. ovaricae;
5: a. ovarica. ϕ 0.5–1.0 mm, 80–100 mm length; *6*: arcade artery. ϕ 0.5–1.0 mm, 80–100 mm

tal direction in the labyrinth plate where their main ramifications spread radially. From these radial main lacunae which now branch in their turn, one central lacuna penetrates each lobe, in the direction of the decidua as well as towards the fetus (Fig. 5b, 7). Since 60 to 80 labyrinth lobes emerge from the labyrinth plate on the decidual side (Fig. 11b), an equal number of arterial central lacunae may be assumed. There are fewer labyrinth lobes pointing towards the fetus. These show only very insignificant arterial central lacunae due to their shortness. Corresponding to what has been described above for the labyrinth lobes as a whole, the arterial central lacunae may anastomose when lobes fuse, or else branch with the cleavage of lobes.

5.2. Maternal Venous Drainage

After the maternal arterial blood has left the capillary lacunae of the labyrinth running from the centre of the lobe to the periphery, it flows into the interlobium (Fig. 8a). The venous blood coming out of the lacunal network of the interlobium (Fig. 6d) accumulates in lacunae of increasing size which finally flow together from the outer two thirds of the main placenta in 20 to 30 large primary marginal lacunae (Fig. 6c). These primary marginal lacunae take their origin in the upper side of the placenta and run around the margin of the placenta (Fig. 6b) radially to the placental stem where they come together to form a basal venous lacunal ring lying in the region where main placenta and stem join (Fig. 6b, c). Its diameter equals approximately that of the placental stem. The basal lacunal ring was distinct in almost all cases examined by us, but varied markedly in its caliber locally. 3 to 5 venous lacunae which not only drain the central labyrinth lobes but also the central third of the fetal surface of the placenta originate in the central region of the main placenta and empty into the lacunal ring (Fig. 6b). Primary marginal lacunae and central drainage lacunae show anastomoses. It is striking that they never incorporate the smallest lacunae of the surrounding interlobar and marginal syncytium but only bigger collective lacunae, although they have no proper vessel walls.

The basal venous lacunal ring has a peculiar histological arrangement corresponding to its localization. Where it is entirely surrounded by interlobium, its wall is formed by interlobial syncytium. This ring may also be encompassed by tissue from the junctional zone in spots where a typical venal wall takes the place of the syncytium. For most of its length, the ring lies at the border of the interlobium and junctional zone. There it is remarkable that the wall in the fetal half is formed by syncytium which will be substituted by endothelium and smooth muscle cells towards the basal side.

The basal venous lacunal ring bulges in two to four spots where the intramural veins branch off (Fig. 6c). Like the spiral arteries the latter follow the margin of the subplacenta around which they are arranged evenly. They reach the uterine wall via the placental stem and thereby incorporate the small subplacental veins, three in average (Fig. 6b). Unlike the arteries, the intramural veins may well show major anastomoses. Hav-

---

length; 7: a. uterina. ϕ 1.0 mm; 8: radial arteries. ϕ 0.5–1.5 mm, 30–40 mm length; 9: intramural arteries. ϕ 1.0mm, 2–4 per placenta; 10: ascending main lacunae. ϕ 0.3–1.0 mm, 4–6 per placenta; 11: radial main lacunae (in the labyrinth plate); 12: lobar central lacunae (in the centre of all lobes); 13: subplacental arteries

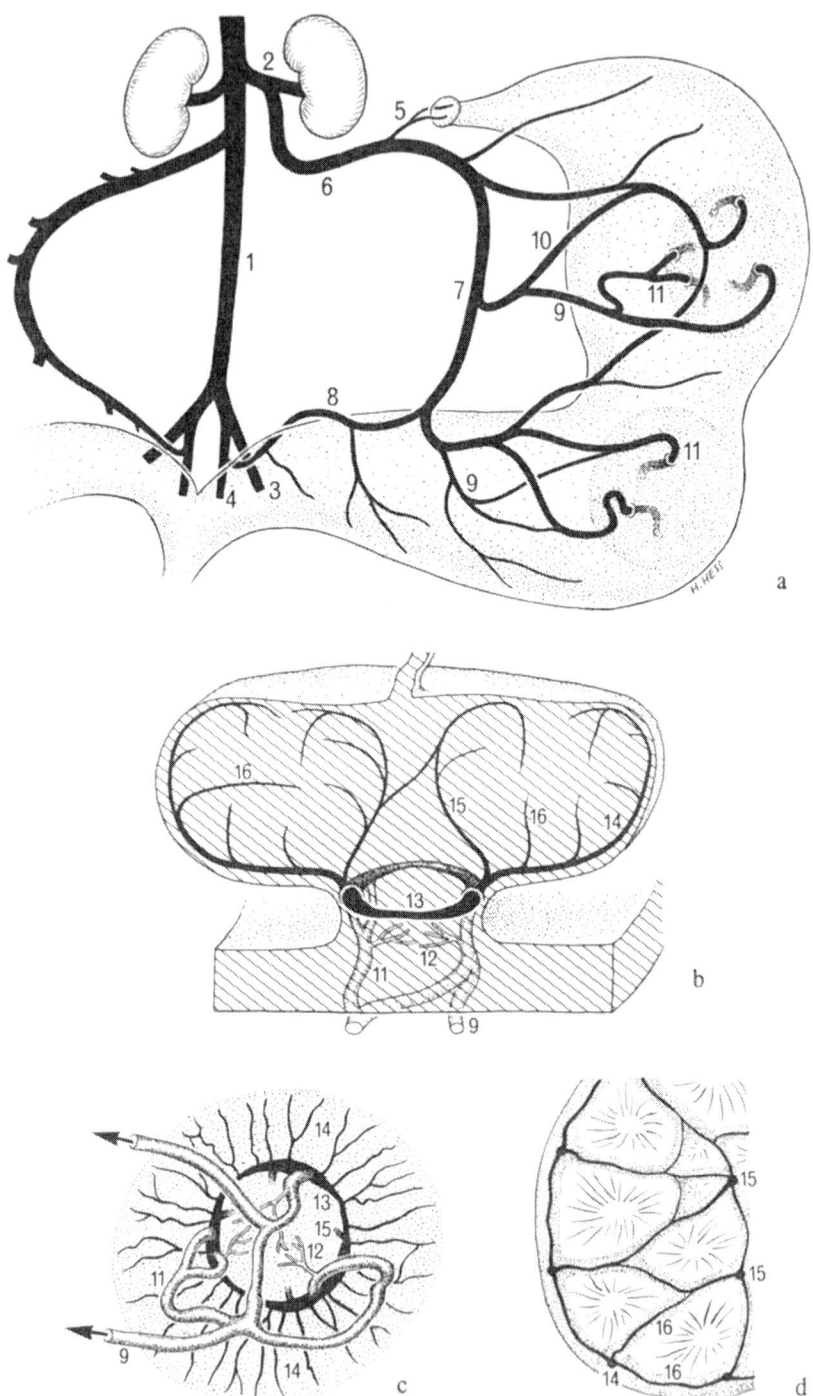

Fig. 6a–d

ing crossed the uterine wall, the veins leave the uterus as radial veins, run through the mesometrium and empty into the arcade vein (Fig. 6a). They correspond to the radial arteries in number and mode of branching, but they have more important cross connections.

Analogous to the arcade arteries, the arcade vein has outlets in the form of the ovarian and uterine vein, but contrary to the arterial conditions, the ovarian vein is the markedly bigger vessel in this case.

5.3. Fetal Arterial Blood Supply

Two fetal arteries enter the placenta via the umbilical cord (Fig. 7). On reaching the main placenta they have anastomoses in most cases and then ramify to form 4 to 6 big arteries lying superficially under the placenta's amniotic cover. A branch originating at the fusion point of the two umbilical arteries traverses the central mesenchymal axis in extension of the umbilical cord. At the level of the subplacenta (or even earlier) it ramifies further. The 4 to 6 superficial arteries usually bifurcate dichotomously. Then the arteries bend to penetrate deeper into the placenta between the lobes. Their dichotomous branching continues while they run along the lobes where interlobium and labyrinth border (Fig. 7). After 10 to 12 bifurcations counting from the umbilical cord, the fetal arterioles plunge radially into the lobes and form a close-meshed capillary network (Fig. 8b).

In very rare cases it may happen that the two umbilical arteries do not anastomose, in which case the areas they supply are independant of each other and differ considerably in size.

5.4. Fetal Venous Drainage

The fetal blood coming from the lobar surface passes through a capillary net, accumulates in the lobar centre and then flows via central lobar veins towards the labyrinth plate (Fig. 8b). The central lobar veins fuse in the labyrinth plate and run radially in

---

Fig. 6a–d. Schematic, largely true-to-scale representation of the maternal veins in the second half of gestation. (a) draining maternal venous system x1, (b) vertical section through the placenta with intramural veins and venous main lacunae. x3. (c) basal view of the main placenta with intramural veins and basal venous lacunal ring. x2. (d) sector out of a horizontal section through the placenta. x4.

1: Vena cava inferior; 2: V. renalis; 3: V. iliaca externa; 4: V. iliaca interna; 5: Ramus ovaricus v. ovaricae; 6: V. ovarica. ϕ 2.5–3.5 mm, 80–110 mm length; 7: arcade vein. ϕ 2.5–3.5 mm, 80–110 mm length; 8: V. uterina. ϕ 0.5–1.0 mm, 80–110 mm length; 9: radial veins. ϕ 1.0–1.8 mm, 30–40 mm length; 10: mesometrial cross-connection; 11: intramural vein. ϕ 1.0–1.8 mm, 10 mm length; 12: subplacental vein. ϕ app. 0.2 mm; 13: basal venous lacunal ring. ϕ (near the draining openings) 0.5–1.0 mm, ϕ (between) 0.1–0.5 mm; 14: primary marginal lacunae ϕ 0.1–0.5 mm, 20–30 per placenta; 15: primary central draining lacunae, 3–5 per placenta; 16: secondary interlobar and marginal lacunae

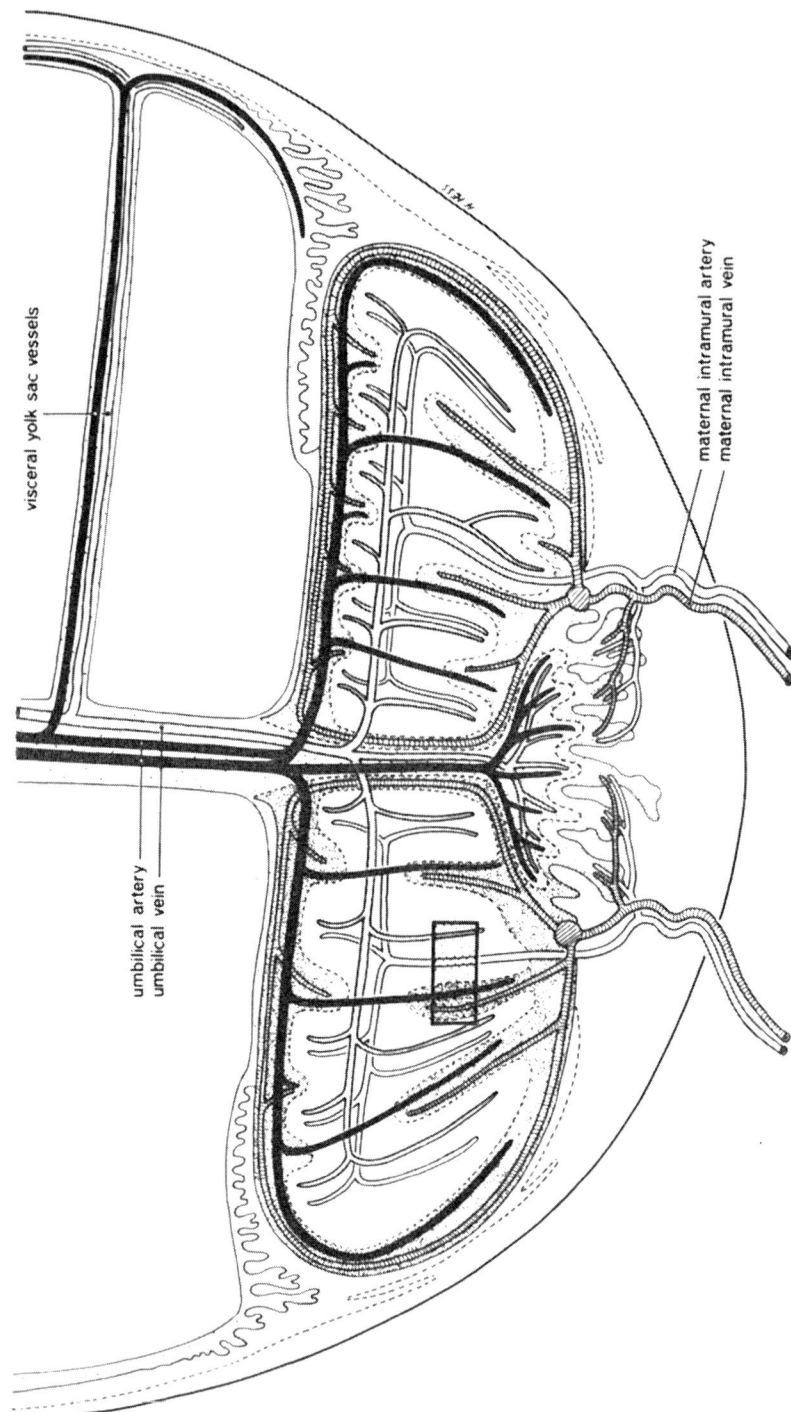

Fig. 7. Synoptic representation of the direction which fetal and maternal vessels and lacunae take. Whereas maternal arterial lacunae and fetal veins run together in the lobar centre, maternal venous lacunae and fetal arteries run in the interlobium parallel to the surface of the lobes. ▭ : section corresponding to the scheme of lacunae and capillaries in Fig. 8

the direction of the central mesenchymal axis where they merge to form the umbilical vein (Fig. 7). In contrast to the arteries, the veins which branch horizontally in star shape do not lie immediately under the surface of the main placenta covered by amnion, but deeper within the labyrinth plate. Here they incorporate the subplacental veins ascending in the central mesenchymal axis from the subplacenta.

5.5. Intralobar Circulatory Conditions

The arterial blood lacunae in the lobar centre (central lacunae) together with the venous blood lacunae in the interlobar syncytium constitute a three-dimensional lacunal network which serves the exchange of substances between mother and fetus (Fig. 8a). In the centre of the lobe, the lacunae still have large lumina and thick walls. On reaching the mid-region of the lobar periphery, the lumina become very narrow and the

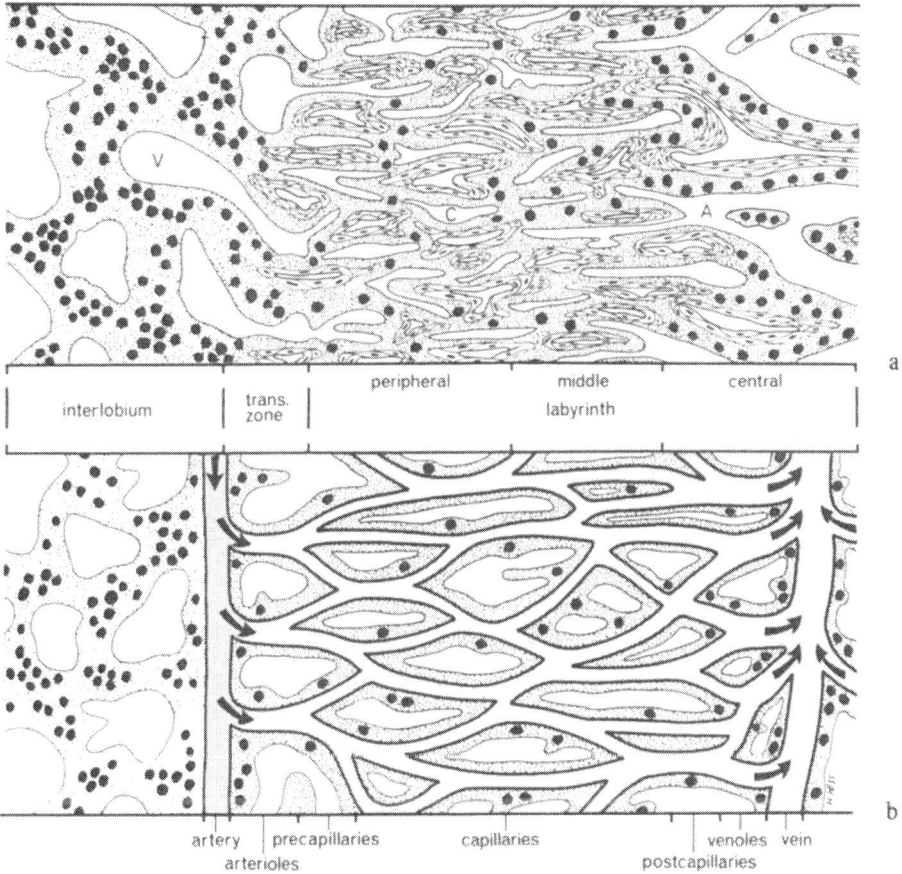

Fig. 8a and b. Greatly simplified representation of the course of the lacunae and capillaries in the interlobium and labyrinth. (a) simplified section view with arterial lacunae (A), capillary lacunae (C) and venous lacunae (V) and in between fetal vessels (b) simplified section view with continuously drawn fetal vessel system

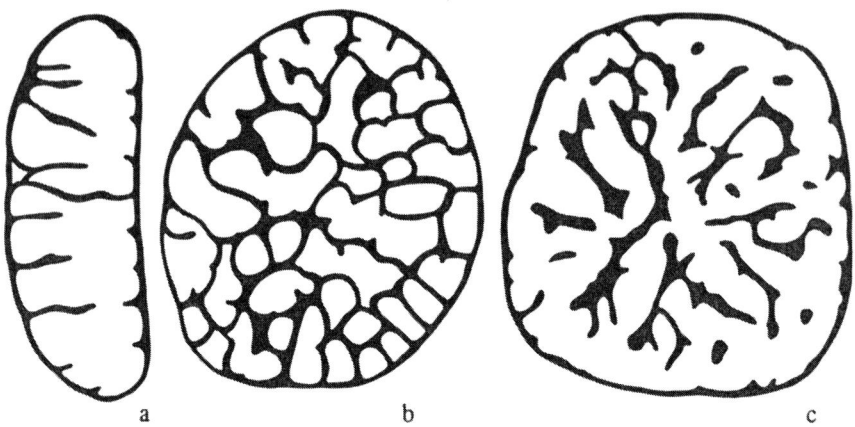

Fig. 9a–c. A true-to-scale drawing of the distribution of interlobium (black) and labyrinth (white) in the mature main placenta of the guinea-pig. (a) vertical section immediately next to the insertion of the umbilical cord. (b) horizontal section at the transition from the middle to the lower third of the placenta. (c) horizontal section at the transition from the upper to the middle third of the placenta. (a) shows short labyrinth cones towards the fetus (on the right), long labyrinth cones towards the uterus (on the left), in between the junction site, the labyrinth plate. It is interspersed by chiefly radial septae of interlobium (horizontal section through the labyrinth plate = Fig. 9c). Fig. 9b shows the long labyrinth cones towards the uterus in a cross-section. All three drawings were made true to the original from the same placenta. X 2

walls may become as thin as $1/4\ \mu$. Entering the interlobar syncytium, the lacunae join again into a system with wide lumina and thick walls which is connected with the interlobar collective lacunae. The spongy syncytotrophoblastic tissue left between the lacunae is also perforated by a three dimensional network of fetal capillaries. They arise from short arterioles at the margin of the lobe and fuse in the centre of the lobe to form big venoles. (Fig. 8b)

In addition to the labyrinthine capillary net, arteriovenous shunts seem to occur between fetal interlobar arteries and collective veins. Ink injection into the fetal circulatory system frequently shows that the supplying and draining vessels fill whereas the capillary net in between remains unstained. The direct indentification of these arteriovenous shunts has not been established so far.

Earlier physiological studies (Leichtweiß and Schröder, 1971) which seemed to confirm their existence, were refuted by the same authors most recently (Leichtweiß and Schröder, 1976): the transfer of tritiated water through the isolated, artificially perfused guinea-pig placenta suggests a non-ideal counter-current arrangement of vessels and lacunae without a functionally significant vascular and diffusional bypass on the fetal side of the exchange area.

It has been known since Mossman's fundamental studies (1926, 1937), that the maternal and fetal blood flows in opposite directions in labyrinthine placentae. In the case of the guinea-pig this means that the maternal blood flows from the centre of the lobe to the periphery, i. e. centrifugally, and the fetal blood centripetally. Hence the morphological and physiological conditions have been referred to as a counter-current system. This view has been contradicted by Müller and Fischer (1968) and by Fischer (1968). These authors discovered in the course of morphological investigations that a lacunal and capillary net with counter-current blood flow intertwine. They interpreted their observations in this way: the maternal blood lacunae come into diffusion contact with several parallel fetal capillaries successively — each capillary comes into contact with a short length of a lacuna only. Following their interpretation, the physiological conditions of a multicapillary system should exist.

We also found two intertwining three-dimensional vessel nets with opposite flow directions. The arrangements of the capillaries, however, should theoretically allow any erythrocyte entering the fetal capillary net to come into contact with the entire length of the intralobular maternal blood lacunae (Kaufmann, 1969a; Bailey, 1974) (Fig. 8a, b). There is no parallel arrangement of the fetal vessels in the sense that an erythrocyte only comes into contact with one section of the blood lacuna. We therefore understand these conditions as a modified counter-current system. Provided that conditions are ideal, this should allow a maximal oxygen saturation of the fetal blood nearly equalling the oxygen content of the maternal arterial blood. If this oxygen pressure is not found in the umbilical vein, as was the case according to the investigations of Bartels et al. (1967) and Fischer (1968), the reasons may be, for example, the thickness of the separating layers, too little contact between the vessels, too big a capillary diameter or too high a flow speed (cp. Dancis et al., 1962). These diffusion impediments should affect CO_2 less as its diffusion capacity is much greater than that of O_2 (cp. Fischer et al., 1965). According to Fischer (1968), the CO_2 content of the blood in the umbilical vein is considerably higher than to be expected in the case of a counter-current system. We therefore consider it possible that a part of the fetal venous blood bypasses the labyrinth via arteriovenous shunts, thus flowing directly from the interlobar arteries into the lobar centre where it mingles with the oxygenized blood. The most recent physiological studies (Leichtweiß and Schröder, 1976) do not attribute functional significance to this eventuality. The authors speak of a nonideal counter-current flow system.

6. Ultrastructure and Functional Morphology

6.1. Physiological, Technical and Morphometrical Data

6.1.1. Physiological Data and Perfusion Technique

Description and evaluation of morphometrical and ultrastructural conditions in the guinea-pig placenta are influenced decisively by the fixation technique chosen. Studies in this field (Kaiser and Kaufmann, 1976) have revealed that perfusion fixation only provides reliable results for both ultrastructure and morphometry (cp. Fig. 13a—f). The most practicable techniques will therefore be presented briefly: Pregnant guinea-pigs are anaesthesized with 60 mg Nembutal per kg. bodyweight, injected in two portions into the muscular system of both hind legs. To avoid reflex movements and disturbance during preparation, local anaesthesia of the abdominal wall has proved to be useful. Moreover, we drip 1 to 2 ml of local anaesthetic into the opened abdominal cavity. Both uterus horns are lifted out of the abdomen and laid laterally so that mesometrium and vessels are tense. To avoid desiccation of the uterus wall, it is covered by gauze soaked in Ringer's solution.

For the *perfusion of the maternal vessel system,* the mesometrium is lifted by placing a gauze pad under it. The uterus is ligated firmly at both ends of the embryo belonging to the placenta to be perfused, in order to prevent an inflow of blood from the neighbouring tissue. With a pincette, a radial artery is caught by the adventitial mantle at its proximal end and stretched. The binding of plastic tubes or catheters is not advisible because of the unavoidable ischemic period. Instead we insert a sharp cannula 1/2 cm deep into the vessel and fix it by hand. Simultaneously the perfusion is started and then all the radial veins which fill with the perfusion liquid are opened. Only then the non-perfused radial arteries leading to the isolated uterus segment are pinched with very small clamps. In this way the arterial supply of the placenta is maintained as long as possible. Ligation of these arteries is necessary to avoid mingling of blood and fixative over a long period of time and thus coagulation. Rinsing of the placenta with an isotone fluid free of fixative beforehand has not proved successful. We merely suck up 0.5 ml of isotone NaCl solution with some heparine after the tube system and the cannula have been filled with fixative. This serves as a barrier between fixative and blood which prevents both mingling and coagulating.

The data below are given as guidance for perfusion pressure and volume: Moll and Künzel (1973) state that the arterial blood pressure in the distal part of the radial arteries in the guinea-pig uterus is 44 mm Hg ($\pm$ 7) in the non-pregnant state and 12 mm Hg ($\pm$ 3) in pregnancy at term. The maternal blood flow rate of placenta and uterus is tabulated below (Table 6).

Since fixation and the low temperature (+ 4° C) influence the vessel caliber, it is impossible to keep both pressure and volume within physiological range during fixation. One has to choose between these alternatives. According to our experience so far it hardly matters whether the perfusion is carried out isobarically or isovolumetrically, figuratively speaking. We were successful in applying glutaraldehyde in a constant flow of 3.75 ml/min per placenta as a standard method for mature placentas. At the beginning of the perfusion, pressure rates of 6 mm Hg are found in the tube system at the cannula; they rise to 16 mm Hg one to two minutes later, and stabilize between 26 and 30 mm Hg after three minutes. Perfusion with OsO_4 leads to simular pressure develop-

Table 6. The maternal blood flow rate of placenta and uterus (according to Bjellin et al., 1975)

Age in days	Main placenta	Subplacenta and junctional zone	Pregnant uterus as a whole
22	0.4 ml/min		
25	0.6 "	0.06 ml/min	0.475 ml/min x g
30	1.1 "	0.12 "	0.385 "
35	2.2 "	0.19 "	0.320 "
40	3.6 "	0.22 "	0.260 "
45	5.0 "	0.20 "	0.210 "
50	6.3 "	0.14 "	0.165 "
55	7.0 "	0.09 "	0.130 "
60	7.3 "	0.08 "	0.100 "
65	7.5 "	0.08 "	0.080 "

The same authors give the mean rates as:

Age in days	Main placenta	Subplacenta and junctional zone	Uterus alone
26–35	0.77 ml/min · g	0.15 ml/min · g	0.32 ml/min · g
56–65	1.53 "	0.15 "	0.18 "

ment at flow rates of 1.5 to 2 ml/min. If these rates are kept throughout and a slightly hypertone fixative is used (2.2 % glutaraldehyde, 340 mosmol, see chapter 6.1.2.) volume relations in the placenta are obtained which are in accordance with physiological measurements (Schröder, 1975). The uniformity of fixation and the ultrastructural preservation were equally good, or even better, and the volume relations the same when perfusion was effected through the fetal vessel system (see chapter 6.1.2.). Physiological data providing guidance for the fetal perfusion are very rare. Shepherd and Whelan (1951) report for a fetal weight of 70 g a blood flow volume of 3 to 4 ml/min in the fetal vessel net. We did not find any data on blood pressure in the umbilical vessels of the guinea-pig. Faber and Hart (1966) give 16 mm Hg for the blood pressure in the umbilical arteries of the rabbit. When perfusing 1 to 2 ml/min slightly hypertone glutaraldehyde (approx. 340 mosmol), the placental volume relations we found were the same as obtained after maternal perfusion and as stated by physiologists (Schröder, 1975). The indicated flow volume for the fetal perfusion leads to a pressure of 20 to 30 mm Hg in the tube system at the cannula a few minutes after perfusion had been started.

The technical proceeding is as follows: After the uterus has been exposed and opened, the fetus is taken out. A small gauze pad is placed under the taught umbilical cord. A sharp cannula is inserted into one umbilical artery and held with a pincette. The cannula is fixed manually and the perfusion started immediately. Here as well a drop of isotone NaCl solution with heparine is put in front of the glutaraldehyde fixative. The umbilical vein is then opened and the umbilical cord ligated between cannula and fetus.

Should a perfusion fixation be impossible or unsuitable for any reason, one can resort to an immersion fixation in glutaraldehyde. However, one then has to reckon with considerable structural deficiency (see chapter 6.1.2.). A 6 % phosphate-buffered glutaraldehyde solution with an osmolarity of approximately 750 mosmol has proved to be a suitable immersion medium for small slices of tissue (not thicker than 1 mm !).

Subsequent to the maternal and fetal glutaraldehyde perfusion which lasts approximately 10 min., slices of tissue not exceeding 3 by 5 by 1 mm are cut out of well perfused areas and postfixed in the same glutaraldehyde solution for 1 1/2 hours. This is followed by postfixation in 1 % phosphatebuffered OsO_4-solution lasting another 1 1/2 hours. Dehydration was achieved by a graded series of chilled alcohol without preceeding rinsing. The samples were embedded in Epon.

6. 1. 2. Influence of Fixation on the Structure of the Placenta

The standard publications on light and electron microscopy of the guinea-pig placenta have been based up to this date exclusively on immersion fixation of the organ (Müller and Fischer, 1968; King and Enders, 1970b, 1971; Enders, 1965; Kaufmann, 1969a; Davidoff and Schiebler, 1970a, b; Davidoff, 1973; Kaufmann et al., 1974). That this method rendered a minor quality of the preservation of organelles in comparison with perfusion fixation should have been known to most of these authors, but was probably considered acceptable even so. It has been known since the investigations of, for example, Pease (1960), Tahmisian (1964), Maunsbach (1966 a, b) and Fawcett (1973), that an adequate perfusion fixation may prevent, in the case of immersion, the unavoidable degeneration of the cytoplasm (lightmicroscopically), dilatation of all organells surrounded by membranes (Fig. 17) as well as lipid extraction, and at the same time may provide an evenly electron-dense image of the plasma. It was certainly not clear, however, that in addition extensive morphometrical alterations were brought about by immersion fixation, which caused an altogether wrong idea of the structure of this organ.

Comparing the intraplacental volume relations, the differences between immersion and perfusion fixation are most striking (Tables 8–11). In the interlobium, for example, the proportionate volume of the lacunae is reduced by about 50 % after immersion compared with the largely natural conditions after perfusion, which means that the interlobar syncytium swells at the cost of the lacunae (Fig. 13a, b). In the periphery of the lobe, such a swelling is naturally less pronounced due to the much thinner syncytial lamellae (Fig. 13c, d). Since the syncytial lamellae also shrink longitudinally (owing to the falling intralacunal pressure), they may increase four to five times in width (Fig. 16b, 17). The result is an artificial image which does not even bear remote resemblance to the natural state as known from the placenta fixed by perfusion.

The thickness of the layers separating the maternal from the fetal blood (Table 11) and the volume relations of maternal and fetal circulation (Tables 8–10) are of decisive importance for all physiological considerations. Their virtually unchanged representation by means of an adequate fixation is therefore not only of academic interest but assumes also practical importance.

For experimental placenta pathology in which the guinea-pig placenta plays an outstanding rôle, the prevention of distorting fixation artefacts is of an equal importance. The edematous swelling of the syncytotrophoblast causing the above mentioned morphometrical alterations is essentially based on a swelling of mitochondria, endoplasmatic reticulum and golgi-system. Just these organelles react most susceptibly to pathological processes. Only a representation free from artefacts allows an exact evaluation of experimental pathology, which can only be guaranteed by an optimal perfusion fixation.

There are several reasons for the superiority of perfusion fixation compared to immersion fixation (Kaiser and Kaufmann, 1976):

1) The maintenance of intravascular pressure in the course of fixation also prevents collapse and form alteration of the lumina and facilitates at the same time the penetration of the fixative into the tissue.

2) The rapid perfusion of the tissue with the fixative guarantees a constant concentration gradient.

3) As the blood circulation is immediately substituted by fixation, an ischemic phase which may alter the ultrastructure can be avoided.

All regions of the main placenta of the guinea pig (interlobium, peripheral and central labyrinth) are vascularized maternally. Since only the labyrinth additionally contains fetal vessels, the choice of the maternal vascular system for perfusion seemed most reasonable. Very good results were obtained locally by this method, which were, however, always accompanied by insufficiently perfused regions showing bad preservation. This disadvantage is not peculiar to perfusion fixation through the fetal vascular system. Moreover this method renders better results than maternal perfusion not only in the labyrinth but even in the interlobium which is not vascularized fetally. This astonishing result may be explained as follows:

1) The fetal vessel system is a clearly arranged terminal vessel system, whereas the maternal lacunal system is a diffuse network of channels of varying width in which a constant flow direction is not necessarily defined by the morphology. Especially in the unphysiological condition of perfusion fixation, blockage of individual lacunae and thus uneven perfusion are easily conceivable.

2) Even more importance may be attached to another fact: while the fixative runs through the fetal vessels the maternal circulation and thus the supply of the placenta with oxygen and nutritive substances is maintained until the fixative has reached the lacunae via the separating layers of tissue (endothelium and syncytotrophoblast) which are then already fixed. Whereas nutrition is replaced by fixation in the case of maternal perfusion, both processes take place simultaneously during fetal perfusion. This is the only possible explanation for the optimal fixation of the interlobium which is not vascularized fetally: its nutrition is also maintained until it is reached – over a long diffusion distance – by the fixative. Since the fetal circulation has no nutritional function for the placenta, these considerations are not applicable to maternal perfusion fixation.

Apart from the flow of the fixative through the organ, the type of fixative and its osmolarity play a decisive rôle in the preservation of structure. Changes in osmolarity of the fixative are reflected nearly linearly by the intraplacental volume relations. The higher the osmotic pressure, the stronger is the dehydration of the tissue. Concomitantly its stainability and electron-density increase. The size of all organelles decreases. All lumina of the organ widen.

Investigations by Kaiser and Kaufmann (1976) show that not only the osmolarity of the fixative is to be considered. During diffusion of the fixative into the tissue the bigger molecules remain behind while the smaller ones diffuse more quickly. Accordingly, the osmolarity at the diffusion front decreases with the distance covered and finally approaches a constant value. Therefore the fixative should be composed more or less hypertonically according to the length of the diffusion distance to ensure an isotonic fixation of the major part of the sample. In the case of relative short diffusion distances in perfusion fixation a slight hyperosmolarity is sufficient. In our experiments values of around 340 mosmol have proved apt. Regarding the longer diffusion distances in immersion we used a 750 mosmol glutaraldehyde solution in order to compensate the drop in osmolarity. It causes a strong shrivelling at the surface of the sample, as from a depth of 40 to 80 μ, however, largely isotonic conditions prevail. We consider the lack of perfusion pressure and the unavoidable ischemic phase responsible for the fact that the structure of the tissue is not well preserved in this depth either.

Regardless of the method of fixation applied, we find in the syncytium of the main placenta of the guinea pig protrusions of the plasmalemm into the lacunae, poor in organelles and filled with plasma (Kaufmann, 1969b). They occur in large numbers in the interlobium (Fig. 13b). For the midgut epithelium Brunings and de Priester (1971) demonstrated the artificial genesis of protrusions caused by inadequate fixation. The influence of fixation is also undoubtable in the guinea-pig placenta (Table 9). They are rare in the case of slightly hypertonic maternal and fetal perfusion fixation (Fig. 13a), and frequent after hypotonic perfusion as well as after immersion (Fig. 13b). We were able to prove experimentally (Kaufmann et al., 1974; Thorn et al., 1974; Kaufmann, 1975a, b; Schneider and Kaufmann, 1976; Thorn et al., 1976) that these protrusions always occur – as a kind of valve – when the intrasyncytoplasmatic osmotic pressure exceeds that in the maternal lacunae. These conditions are given for example when perfusing hypotonically, but may also occur following a rise of the intrasyncytoplasmic osmotic pressure caused by blockage of metabolism (Thorn et al., 1974). The formation of these plasmal protrusions after immersion fixation is presumably due to an intrasyncytial rise of the osmotic pressure caused by ischemia.

Similar plasmal protrusions occur in great numbers in vivo in the placenta patholog-ically and play a major part in infarct genesis (Stark and Kaufmann, 1974; Kaufmann, 1975b). Exclusion of artifical formation (e. g. inadequate mode of fixation, wrong os-molarity) is of particular importance for their analysis. As the protrusions are either washed away during maternal perfusion or else hinder fixation by blocking lacunae, fetal perfusion fixation is most suitable here.

The type of fixative (glutaraldehyde and/or osmiumtetroxide) has a decisive influ-ence on stainability of the tissue, on electron-microscopical contrast and depiction of membranes (see Sabatini et al., 1962, 1963; Schiechl, 1971). The finestructural and light-microscopical volume relations, however, are hardly affected.

Fixation with glutaraldehyde alone is not suitable for finestructural morphology be-cause of deficient contrast of membranes. Sole OsO_4-fixation, on the other hand, is characterized by sharp-lined structures. This is caused by only faint representation of plasmatic elements and by well contrasted membranes. According to Dallam (1957) this phenomenon is due to considerable loss of protein in the plasma during fixation. Luft and Wood (1963) contradicted this interpretation. Irrespective of the explanation of this phenomenon, sole OsO_4-fixation is suitable for very electron-dense tissue which are then easier to survey. The placental syncytotrophoblast counts among these tissues. Owing to poor diffusion rates (Palade, 1952; Caulfield, 1957), the applicability of OsO_4 is restricted: an immersion fixation without previous glutaraldehyde fixation is only acceptable for samples of tissue of 1/10 to 1/4 mm diameter. Such small samples only allow a limited survey and so this proceeding seems unadvisable for the guinea-pig placenta which is composed of very defined regions. Even perfusion-fixation with OsO_4 renders a worse preservation of organelles than glutaraldehyde perfusion. Mauns-bach (1966a, b) confirms this for the kidney also. The bad diffusion properties should be held responsible here as well.

Double-fixation with the rapidly penetrating glutaraldehyde and subsequently with OsO_4 to contrast the membranes seems to be the best solution in nearly all aspects. It ensures an optimal representation of structure and is advantageous for comparison since it is the most widely applied method of fixation. As far as experimental conditions and aims allow, we perfuse via the fetal vessel system with glutaraldehyde and add an immersion fixation with OsO_4 (see chapter 6.1.1.).

6. 1. 3. Morphometry of the Main Placenta

As far as we know, there are only three publications on the guinea pig placenta in which morphometrical data are given. Ibsen (1928) published data on weight develop-ment of the placenta and of the fetal membranes, junctional zone and subplacenta (see Table 4). Müller and Fischer (1968) analyzed the fetal and maternal blood circulation in the guinea-pig placenta, and stated the contact surface between the two to be 0.30 m^2. In one of our own publications (Kaufmann, 1969a) there are some estimated data on the share of interlobium, periphery of the lobe and centre of the lobe in the placenta during its development. However, there are no precise data up to now on the volume relations of the different tissues in the guinea-pig placenta, on the volume of the maternal and fetal blood spaces, on the thickness of the separating layers between maternal and fetal circulation and on vessel surface involved in materno-fetal and feto-maternal exchange.

In attempting to fill this gap in the past few years, we encountered such big problems, that we have not yet been able to put forward statistically proved figures which seemed reliable enough to us. The following figures on the morphometry of the main placenta are only provisional approximations which may serve as a rough orientation for the time being. A critical discussion of these data, a possible correction as well as statistical control will have to be left to a future publication (Stelter and Kaufmann, 1976). Some data giving orientation on the morphometry of the subplacenta will be dealt with in chapter 6.4. On the morphometry of the yolk sac placenta we have no data whatsoever.

The crucial problems in the morphometry of the guinea-pig main placenta are 1. the very bizarre distribution of its manifold components, which is not without regularity in terms of statistics and therefore difficult to calculate, and 2. the strong dependance of all data on the way of fixation. The second factor in particular renders reliable statements more difficult: the lacking uniformity of this organ demands large survey sections for evaluation, i. e. paraffin-sections with all the disadvantages of deficient structural preservation and uncontrollable shrinking. The in these aspects better technique of the glutaraldehyde-fixation with epoxy-resin embedding does not provide a sufficient survey thereby posing statistical problems. We therefore tried to combine the two methods.

The mode of fixation (immersion or perfusion), the osmolarity of the fixative and the perfusion pressure are further important variables which have a strong influence on

Table 7. Interpolated figures on the volume development of the rough tissue components in the guinea-pig main placenta without the parietal yolk sac cover after immersion fixation according to Bouin and embedding in paraffin. The border between interlobium and labyrinth was drawn at the margin of capillarization. The border between periphery of the lobe and centre of the lobe was set where the mean luminal width of lacunae and venoles exceeded 40 μ. The mean weights of the main placenta (in grammes) (without parietal yolk sac cover, subplacenta and junctional zone) were won by using the weight data of Ibsen (1928) as well (cp. table 4)

Age in days	Interlobium including big vessels	Labyrinth including big vessels	Periphery of the lobe	Centre of the lobe including big vessels	Mean weight of the main placenta
12	100 %	0 %	0 %	0 %	
15	77	23	17	6	
18	66	34	25	9	
21	58	42	31	11	0.1
24	52	48	37	11	0.2
27	46	54	43	11	0.4
30	41	59	48	11	0.6
33	37	63	52	11	0.8
36	33	67	56	11	1.3
39	30	70	59	11	1.8
42	27	73	62	11	2.4
45	25	75	64	11	3.0
48	23	77	66	11	3.4
51	21	79	68	11	3.9
54	19	81	70	11	4.4
57	17	83	72	11	4.7
60	16	84	73	11	4.9
63	15	85	74	11	4.8

all values. As pointed out in the preceeding chapters, we consider perfusion fixation through the fetal vascular system with 2.2 % glutaraldehyde (340 mosmol) to be optimal as to ultrastructural preservation and morphometrical evaluation. We perfused at a constant flow volume and recorded the perfusion pressure in the tube system in front of the cannula (luminal width 0.90 mm) whose tip had been inserted in one of the umbilical arteries. At pressures of 20–30 mm Hg we obtained the values which seemed most reliable according to physiological comparative data (Schröder, 1975). As we cannot yet vouch for the correctness of these data, we opposed them in most cases to data won after perfusion fixation with the same solution but a different pressures, as well as to data obtained after immersion fixation.

The data on development of the volume of the interlobium, of the periphery of the lobe and of the centre of the lobe (Table 7) are based on paraffin-embedded samples after Bouin immersion fixation. Comparison with epoxy-resin embedded material after perfusion fixation reveals no considerable deviation in this case. The differences to our earlier estimated data (Kaufmann, 1969a) are mainly due to newly defined borders between the different regions which are rather indistinctly deliminated. The most striking findings is, as already stated in an earlier chapter (3.3.), the astonishing constancy of the volume of the central labyrinth in the main placenta. For better comprehension, the percentual data were complemented by the absolute main-placental weights (fetal membranes, subplacenta and junctional zone having been substracted).

The morphometrical analysis of the labyrinth (Table 8) was carried out on semi-thin sections. Most obvious was the enormous increase in volume of the lacunae with

Table 8. Morphometrical analysis of the labyrinth of the mature guinea-pig placenta without the big supplying and draining vessels (over 100 μ diameter). The figures in the boldly rimmed column which were won after perfusion fixation of the fetal vessels at a pressure of 20–30 mm Hg in the tube system in front of the cannula, appear to approximate the in-vivo relations best

	Immersion fixation with 6 % glutaraldehyde (750 mOsmol)	Perfusion fixation of the fetal vessels with 2.2 % glutaraldehyde (340 mOsmol) perfusion pressure			
		<20mmHg	20–30mmHg	31–50mmHg	51–80mmHg
Percentual volume of fetal capillary lumina	19 %	14 %	17 %	15 %	13 %
Percentual volume of maternal lacunae	11 %	44 %	35 %	42 %	49 %
Percentual volume of the trophoblast	70 %	42 %	33 % ⎱ 48 % ⎰ 15 %	43 %	38 %
Percentual volume of endothelium and connective tissue					
Fetal capillary surface per mm³ of labyrinth	120 mm²	106 mm²	148 mm²	127 mm²	94 mm²
Maternal lacunal surface per mm³ of labyrinth	131 mm²	141 mm²	152 mm²	152 mm²	154 mm²

concomitant decrease of trophoblast and endothelium after perfusion in contrast to immersion. The variations of the capillary volumes, on the other hand, are much less conspicuous and probably not even statistically significant. Under optimal conditions of fixation, the lumina ought to be in a relation of 1:1 to the trophoblast together with the endothelium and in both, the capillaries to the lacunae and the endothelium to the trophoblast in a relation of 1:2 respectively.

An important filtration of water from the fetal into the maternal vascular system is evidently to be held responsible for the seemingly contradictory finding that an increase in perfusion pressure in the fetal vascular system is followed by dilatation of the maternal lacunae but not of the capillaries. Vice-versa, this result is not applicable to maternal perfusion fixation. This may be due to a purely morphological factor: the capillaries mostly bulge into lacunae twice their size. With rise of fetal perfusion pressure they are therefore able to filtrate optimally, analogous to the conditions in the glomerulus of the kidney. With increase of pressure in the lacunae, however, the capillaries are compressed. The fluid nevertheless passing through the trophoblast cannot be taken up by the compressed fetal vessels (like in the glomerulus of the kidney, where a re-filtration of urine into the capillaries is impossible even when the pressure rises in the Bowman's capsule!) and accumulates in the intercellular spaces. This is an artefact common to maternal perfusion fixation which we have never observed in fetal perfusion.

The conditions in the interlobium are much clearer (Table 9). Here there are no fluctuations in volume worth mentioning when the perfusion pressure varies between 10 an 90 mm Hg. The distinct difference after immersion fixation appears logical: due to the collapsing of the lacunae, the whole of the interlobium loses volume. The share of the trophoplast grows relatively as well as absolutely due to hydropical swelling. This causes a concentration of the lacunae, which in turn leads to an increase of their surface area per volume unit. The absence of polypous plasmal protrusions of the trophoblast in the lacunae in the normal placenta is a safe sign for the good quality of the fixation, as shown in Table 9. As pointed out elsewhere (see chapter 6.1.2., Kaufmann,

Table 9. Morphometrical analysis of the interlobium of the mature guinea-pig main placenta without big supplying and draining vessels (over 100 μ diameter). In perfusing the fetal vascular system we did not find significant volume changes at pressures between 10 and 80 mm Hg. The values won after perfusion fixation should approximate the in-vivo relations

	Immersion fixation	Fetal perfusion fixation
Percentual volume of the maternal lacunae in the interlobium	29.0 %	42.1 %
Percentual volume of the trophoblast in the interlobium	71.0 %	57.9 %
Percentual volume of the plasmal protrusions in the lacunae	11.2 %	< 0.1 %
Lacunal surface per mm³ of interlobium	91 mm²	71 mm²

Table 10. Volume relations of the tissue components in the mature guinea-pig main placenta including the parietal yolk sac cover after perfusion fixation of the fetal vascular system with 2.2 % glutaraldehyde (340 mOsmol) at a perfusion pressure of 20–30 mm Hg. Big lacunae and big fetal vessels are those with a diameter of over 100 μ. In boxes: sites of metabolic and exchanges performances in the main placenta. Together they amount to 83.5 % of the volume of the main placenta including the parietal yolk sac cover

Parietal yolk sac epithelium with Reichert's membrane		4–6 %		
Interlobium including big vessels	connective tissue and vessel walls	1 %		14 %
	large lacunal lumina	2–2.5 %		
	large vessel lumina	1.5–2 %		
	remaining interlobium: trophoblast	5.2 %	} 9.0 %	
	lacunal lumina	3.8 %		
	periphery of the lobe: trophoblast	23.1 %		70 %
	vessel walls and connective tissue	10.5 %		
	lacunal lumina	24.5 %		
	capillary lumina	11.9 %		
Labyrinth including big vessels	centre of the lobe: walls of big vessels	0.5–1 %		11 %
	large lacunal lumina	3.5–4 %		
	large fetal vessel lumina	2.0 %		
	remaining centre of the lobe: trophoblast	1.4 %	} 4.5 %	
	vessels walls and connective tissue	0.7 %		
	lacunal lumina	1.6 %		
	capillary lumina	0.8 %		
				100 %

1975b) the protrusions may also occur in large quantities as a manifestation of pathological processes in the trophoblast even under ideal fixation conditions.

The combination of the figures in the Tables 7−9 together with the registration of the big supplying and draining vessels (width of the lumina over 100 μ) gives the compilation in table 10. It facilitates the calculation of the absolute volumes of the various tissue components of the mature guinea-pig placenta for a given total weight. After loosening the necrotical tissue of the junctional zone, the parietal yolk sac usually remains attached to the placenta and can only be completely removed with loss of time and damage of the placental surface. It is therefore comprehended in the data on the main placenta. Hereby, however, it has to be considered that its volume (5 % in the mean) has a greater share in small placentae (3−3.5 g) amounting to approximately 6 %, decreasing to 4 % in big placentae (7−10 g).

The interest of physiologists in anatomical research is mainly focussed on data on inner surfaces and on the thickness of separating layers; this applies to most organs and to the placenta as well. Despite all reservations one may have as to the methods (see above), we would like to give some information concerning the data metioned (Table 11). However, one has to reckon with possibly not unimportant deviations after evaluation of extensive material (Stelter and Kaufmann, 1976).

The mean thickness of the separating layers between maternal and fetal vessel lumina was calculated by dividing the total volume of trophoblast, endothelium, connective tissue, basal membranes and extracellular space by the mean surface of capillaries and lacunae. The relatively small variation of the values between 3.0 and 3.4 μ after perfusion pressures of 10 to 60 mm Hg speaks for their accuracy. The considerable deviations of 5.6 μ (after immersion fixation) and 1.5 μ (after perfusion at pressures of over 60 mm Hg) only occur under conditions which are so definitely unphysiological that these results are beyond discussion.

Using material obtained unter presumably ideal conditions of fixation, we additionally categorized the thickness of the separating layers in 5 classes and calculated their share in the materno-fetal exchange area with the help of electron-micrographs. The sum of these values gives a slightly higher mean diffusion length than stated in the preceding paragraph. This is due to varying angles of the sections in relation to the surface of the vessels which gives a deceptive idea of the actual thickness of the layers. We have not been able to correct these data up to now as the conditions in the labyrinth are difficult to survey spatially as well as mathematically.

Data on the surface of lacunae and capillaries differ considerably according to the method of fixation used. It is noteworthy that the smallest differences between the two resulted at the probably optimal perfusion pressures of 20−30 mm Hg. Under these conditions we obtain a common contact area of approximately 0.11 m² per cm³ main placenta. Comparing these figures with the data of Müller and Fischer (1968), we largely find congruence. These authors give 0.30 m² for a placental volume of 2.71 cm³.

Table 11. Tabular compilation of the exchange surfaces obtained after different ways of fixation (in m² related to 1 cm³ mature guinea-pig main placenta including parietal yolk sac cover) and diffusion lengths from maternal to fetal blood and vice versa. The mean diffusion length was calculated by dividing the volume of the solid tissue components (trophoblast, endothelium, connective tissue) by the mean of capillary and lacunal surface. The classification of the diffusion lengths was obtained by measurements on electron-microscopical survey pictures. The figures in the boldly rimmed column won at perfusion pressures 20–30 mm Hg should approximate the in-vivo relations best

	Immersion fixation	Perfusion fixation of the fetal vascular system with 2.2 % glutaraldehyde (340 mOsmol) at the following perfusion pressures			
		$<$ 20 mm Hg	20–30 mm Hg	31–60 mm Hg	$>$ 60 mm Hg
Maternal lacunal surface per cm³ main placenta — interlobium	0.008 m² ⎱ 0.106m²	0.007 m² ⎱ 0.112m²	0.007 m² ⎱ 0.120m²	0.007 m² ⎱ 0.121m²	0.007 m² ⎱ 0.117m²
labyrinth	0.098 m² ⎰	0.105 m² ⎰	0.113 m² ⎰	0.114 m² ⎰	0.110 m² ⎰
Capillary surface per cm³ main placenta	0.089 m²	0.078 m²	0.110 m²	0.083 m²	0.078 m²
Mean diffusion length between capillary and lacunal lumina in the periphery of the lobe	5.6 μ	3.4 μ	3.2 μ	3.0 μ	1.5 μ
Classification of the diffusion lenghts between capillary and lacunal lumina in the periphery of the lobe — up to 1.0 μ	./.	./.	16 %	./.	./.
1.1–3.0 μ			38 %		
3.1–5.0 μ			22 %		
5.1–7.0 μ			19 %		
over 7.1 μ			5 %		

6.2. Ultrastructural and Histochemical Development of the Main Placenta

6. 2. 1. 6th to 14th Day

Electron microscope studies by Enders and Schlafke (1969) have shown that the guinea-pig blastocyst before establishing contact with the uterine mucosa (6th gestation day) has a comparatively well developed syncytium, formed as the result of lysis of the cytotrophoblastic cells' marginal membranes. According to Davies et al. (1961b) on the 15th day, the main placenta trophoblast, situated above the subplacenta, is made up of overlapping cytotrophoblastic cells; the latter finding was subsequently but only partially confirmed by Davidoff and Schiebler (1970b) and Davidoff (1973), who established that between the 12th and 14th day of pregnancy, the placenta build-up includes syncytiotrophoblast, cytotrophoblast and mesenchymal cells. Syncytiotrophoblast is the prevalent component of the main placenta in the term referred to. Cytotrophoblastic cells are met with comparatively rarely. A greater number is observed in the peripheral areas of the placenta and in the vicinity of the central excavation. The earliest signs of invasion by mesenchymal cells of the nearby syncytium are recorded in the central excavation zone.

The syncytioplasm in the period under review is endowed with a weak electron density. Its nuclei are situated in groups or in singles, they are comparatively large, and posess irregularly positioned chromatin (Fig. 12). Mitoses are also encountered, but rather rarely. The rough endoplasmic reticulum (reR) differs in quantity in the various zones of the syncytioplasm. Certain zones contain great amounts of free and polyribosomes. Tiny fields with smooth endoplasmic reticulum (seR) are noted very rarely. The mitochondria present variable forms and sizes. Forms with a bright matrix and cristae oriented in different directions prevail. A reduced number of small mitochondria with a dense matrix are also met with, as well as mitochondria containing tubules. The number of Golgi-complexes is high. They are situated chiefly perinuclearly, and some of them are distended. In the vicinity of the Golgi zones, besides the characteristic vesicles, single dense core vesicles with diameter $0.17-0.2\ \mu$ are also discovered. Some of the syncytioplasma fields are filled out with numerous pinocytotic vesicles with average diameter $0.08-0.17\ \mu$. The smallest among them ($0.01\ \mu$) are electron dense and are detached from deep, resembling narrow channels, invaginations in the plasmalemma. Deposition of the reaction product following electron-microscopic demonstration of ATPase activity along their membranes is assumed as evidence ot their derivation

---

Abbreviations used in this and the following chapters:
Acetylcholinesterase = AChE, acid 5-nucleotidase = a5Nase, acid phosphatase = acP, adenosine triphosphatase = ATPase, alkaline phosphatase = alkP, citrate cycle enzymes = CC, cytochrome oxydase = CO, glucose-6-phosphatase = G6Pase, glucose-6-phosphate-dehydrogenase – G6PDH, glutamate-dehydrogenase = GDH, α-glycerophosphate-dehydrogenase = α GPDH, hexokinase = HK, β-hydroxybutyrate-dehydrogenase = βHBDH, hydroxysteroid-dehydrogenase = HSTDH, isocitrate-dehydrogenase = IDH, lactate-dehydrogenase = LDH, leicylaminopeptidase = LAPase, malate-dehydrogenase = MDH, nicotinamide adenine dinucleotide = NAD, nicotinamide adenine dinucleotide phosphate tetrazolium reductase = $NADPH_2$-Red, nicotinamide adenine dinucleotid tetrazolium reductase = $NADH_2$-Red, 6-phosphogluconate-dehydrogenase = 6PGDH, ribonucleic acid = RNA, succinate-dehydrogenase = SDH.

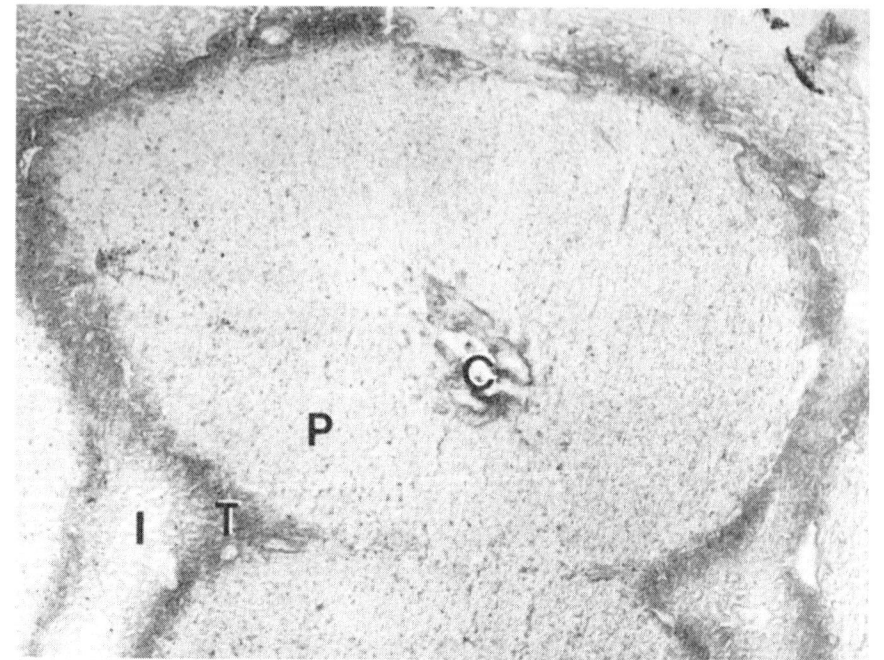

Fig. 10

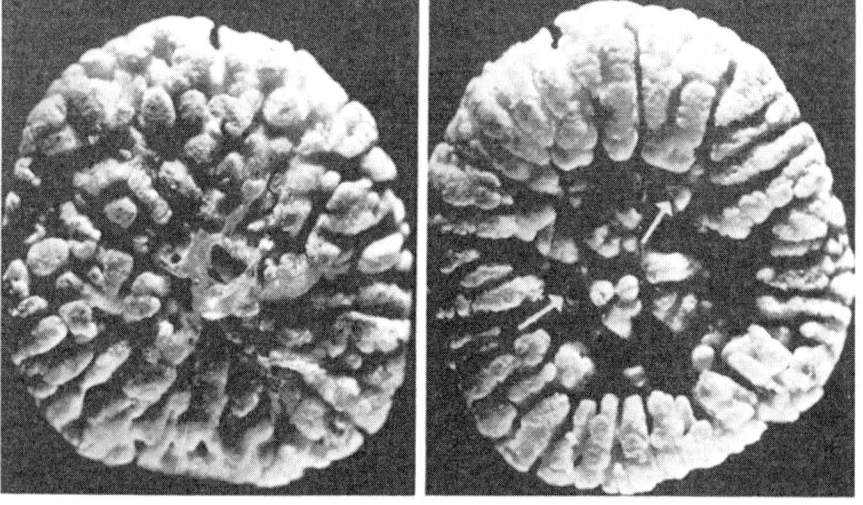

Fig. 11

a   b

Fig. 10. Enzymehistochemical representation of the alkaline phosphatase in the unfixed cryostat section (method: Gomori, 1952). Owing to the omission of prefixation, only the unsoluble, structure-bound portion of the enzyme is depicted. This is nearly exclusively localized in the fetal arteries and arterioles as well as in the fetal veins and venoles. Thus the location of these vessels, the transitional zone (T) and the centre of the lobe (C) are marked. The peripheral labyrinth (P) and the interlobium (I) react weakly or negatively. This method is especially suited for a survey representation of the different parts of the main placenta. X 35

Fig. 11. Plastoid-injection of the fetal arteries and capillaries of a mature guinea-pig placenta. After maceration of the placental tissue we see a casting of the labyrinth. (a) view from the fetal side on the short labyrinth cones and on the labyrinth plate; in the centre the umbilical vessels. (b) basal view on the long labyrinth cones. The central hollow corresponds to the site of the necrotic subplacenta. In the basal tips of some of the middle cones one observes apertures (arrows) which correspond to the entrance of the ascending arterial main lacunae. X 2.5

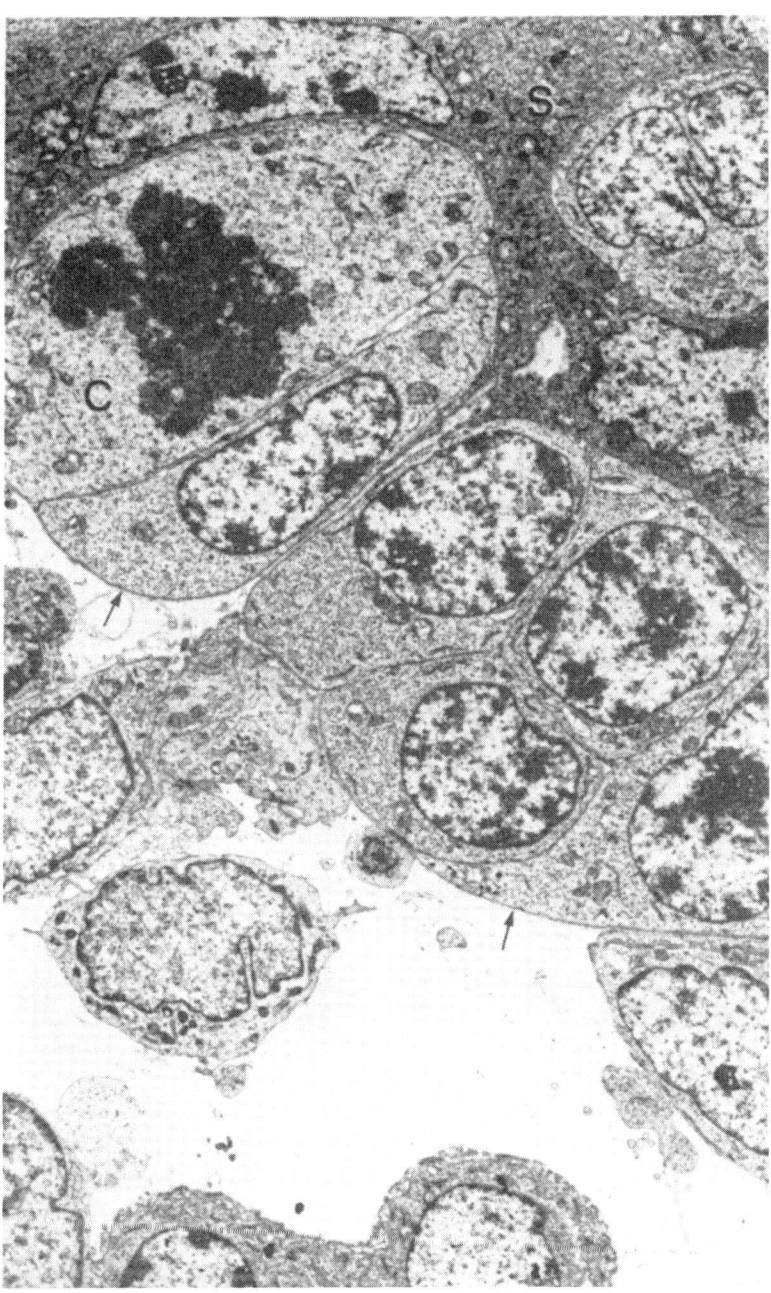

Fig. 12. Micrograph of the guinea-pig placenta from the 14th day. The picture shows the bordering zone between penetrating mesenchyme (below) and interlobium (above). In between a layer of light cells (C) having desmosomal contact with the interlobar syncytotrophoblast (S). One of these cells is in mitosis. Since these cells are separated by a basal lamella (arrows) from the mesenchyme but not from the trophoblast, it is most probably cytotrophoblast. Immersion-fixation with 1 % OsO$_4$, X 5200

from the plasmalemma invaginations. At this particular period, the lysosome-like structures are comparatively abundant. Multivesicular bodies, corpuscles containing vesicles and other structures, dense corpuscles, myelin-like corpuscles, and autophagic vacuoles with diameter 0.13–1.0 μ are encountered. The presence of numerous lipid drops, differing in form and size (0.2–5.7 μ), part of which display direct contact with the mitochondrial membranes, is worth noting. Small lipid droplets are situated between the microvilli of the syncytioplasma which encompasses the well-shaped maternal blood spaces. These lipid droplets enter the syncytioplasma by pinocytosis.

The cytotrophoblastic cells between the 12th and 14th day of gestation are usually positioned in groups or else form rows (Fig. 12). Between their marginal membranes, as well as between the membranes of the cytotrophoblastic cells and syncytiotrophoblast, desmosome contacts are often established. The cytoplasm of these cells is electron bright, contains numerous ribosomes and polyribosomes, few reR, mitochondria and Golgi zones (GA). The cytotrophoblast cell nuclei are large and rich in chromatin. Often, mitoses are also distinguishable (see Wynn, 1968). The marginal membrane of single cytotrophoblast cells is interrupted, and their cytoplasm merges with the syncytioplasm. By its appearance, the cytoplasm of these cells resembles the syncytioplasm. The latter finding is corroborated by the presumption that the syncytioplasma is made up of the cytotrophoblast cytoplasm after rupture of the cell membranes (Enders and Schlaffke, 1969; Kaufmann 1972).

The mesenchymal cells between the 12th and 14th gestation day are situated within the central excavation. Single projections invading the syncytiotrophoblast can be seen. Morphologically, the mesenchymal cells show a great diversity (Fig. 12). Part of them strongly resemble the cytotrophoblastic cells, and their identification proves rather difficult. The cytotrophoblastic cells can be distinguished with certainty from the similar looking mesenchymal cells after the formation of the basement membrane (20th–22nd gestation day) between the mesenchymal projections and syncytioplasma. A basement membrane is never discovered between cytotrophoblast and syncytium. Another part of the mesenchymal cells resemble fibroblasts, and their outgrowths are accompanied by parallely disposed collagen fibers. The other mesenchymal cells disclose a great diversity in terms of form and size, with the number of mitochondria and presence of the remaining cell organelles (seR, ribosomes, GA and the like) varying considerably from cell to cell. The mesenchymal cells seldom form a lumen wherein nucleusbearing fetal erythrocytes are visible. These first fetal "capillaries" are established within 14 days of gestation. Simultaneously with the above processes, autolytic processes also develop in the syncytioplasma, where apparently the syncytioplasmic nuclei play a peculiar role. In the autolytic apertures resulting the penetration is observed of mesenchymal projections which form the vessels.

6. 2. 2. 15th to 20th Day

Between the 15th and 20th day of gestation, an intense proliferation of mesenchymal projections toward the syncytioplasma is noted. During the latter period, it is already possible to distinguish the syncytiotrophoblast containing mesenchyma and a few blood vessels (the future labyrinth), and the trophoblast free of mesenchyma (the future interlobe). The presence of strong distensions of reR is characteristic of the syncytioplasma. Simultaneously, most of the GA undergo intense activation, and dense material begins to accumulate in their cisterns. The secretory granules thus formed break off and migrate throughout the syncytioplasma. The peak of these GA secretory phenomena occurs between the 18th and 20th day, when a large amount of secretory granules are localized on the surface enveloping the maternal blood spaces. Isolated granules come into contact with lipid drops which enter by pinocytosis. To all appearances, the peculiarities described are related to the steroid metabolism of the placenta (Piziak and Gavienowski, 1973). This is furthermore supported by the results of Davidoff and Gospodinov (1971), who succeeded in establishing $3\alpha-$, $3\beta-$, $\Delta^5 3\beta-$, $11\beta-$ and $17\beta-$HSTDH activity by the 18th day, mainly perinuclearly and underneath the maternal blood spaces of the mesenchymafree syncytiotrophoblast (cp. Ferguson and Christie, 1967). The reaction around the free surface of the trophoblast is manifested to a still higher degree by the 20th gestation day when it is also noted at the base of the syncytioplasmic polyps then forming. It is likely that the above results point to the fact that lipid drops (probably containing cholesterol), as well as other substances derived from the maternal blood (see Kayden, 1968; Kayden et al., 1969) play an important rôle in the steroid synthesis within the guinea-pig placenta at this particular period (see Simmer, 1968). Along with intensification of the signs of secretion and steroid metabolism, corroborated both by the GA peculiarities, and by the presence of tubular mitochondria and smooth endoplasmic reticulum, a substantial increase in quantity of autophagic bodies and other lysosomelike structures begins. This is an indirect indication of a general trophoblast metabolism intensification.

In this period, comparatively well pronounced SDH (Reale and Pipino, 1959), CO, NADH$_2$ Red., NADPH$_2$ Red. (Vollrath, 1965), MDH, IDH (Christie, 1968), acP, a5Nase, ATPase (Davidoff and Schiebler, 1970b) and AChE (Goutier-Pirotte and Gerebtzoff, 1955; Gerebtzoff, 1957; Perrotta and Lewis, 1958) are observed. In the junction zones between free and mesenchyma-containing trophoblast (transitional zone) the activity of G6PDH (Vollrath, 1965) alkP (Fig. 10) and hexokinase is rather marked. On their part, the mesenchyma-invaded zones of the syncytiotrophoblast exhibit strong alkP (Vollrath, 1965; Kaufmann, 1974), LAPase, hexokinase (Kaufmann, 1974), G6PDH, LDH, GDH (Vollrath, 1965) and ATPase (Davidoff, 1970). Thus, in the above period, the histochemical results allow us to identify zones with a peculiar metabolic characteristics as early as before the morphological differentiation of the main placenta. The labyrinth in formation is already delineated as a zone of a more active protein metabolism, whereas the capillary-free syncytium — as a zone respresenting the energy station of the placenta. On the other hand, the occurence of a greater number of autophagic vacuoles and various lysosome-like structures, manifested much more strongly in the following placental development stage, could be partly related also to the secretory process regulation in the light of the data, published by Smith and Farquhar (1966), concerning the hypophysis, i. e. the excess quantities of secretory material or the granules carrying the material, after exhaustion, are being incorporated in the lysosome turnover and undergo hydrolytic splitting.

During this period, division of the syncytioplasma by streaks of varying width into zones with different amounts and combinations of the cell organelles is observed. The streaks represent a combination of fine granular substances and of osmiophilic filamentous bundles with different length and width situated within the latter. The peculiarity outlined points to a different functional state in the adjacent syncytioplasmic zones. The number of cytotrophoblastic cells shows a progressive reduction.

6. 2. 3. 21st to 35th Day

Between the 21st and 35th day, a clearcut separation of the placenta into capillary-free interlobar and marginal syncytium (interlobe), and capillarized labyrinth takes place. The nuclei in the interlobe are of diminished sizes, and the number of GA releasing secretory granules decreases. This is in compliance with the strong reduction of HSDH-activity, established by Davidoff and Gospodinov (1971). Simultaneously, the quantity of lipid drops in the syncytioplasm decreases, while the number of mitochondria with a compact matrix increases, which is also valid for the lysosome-like structures. A larger amount of filaments is also met with.

In addition, the labyrinth formation coincides with considerable changes in the histochemical picture of the syncytiotrophoblast of the main guinea-pig placenta. In the interlobe a considerable AChE activity is noted. We succeeded in demonstrating alkP in unfixed cryostat sections, which is negative on fixed material in the particular zone. Well pronounced activity has been described for ATPase (Davidoff, 1970) and for a number of oxidoreductases as well $NADH_2$ Red., $NADPH_2$ Red., GDH, βHBDH, αGPDH, CO, LDH, G6PDH (Vollrath, 1965). The transitional zone to the labyrinth is once again distinguished by the marked activity of alkP, hexokinase and G6PDH.

The strong proliferation of mesenchymal projections from the 22nd gestation day onwards accounts for further distension of the labyrinth. The established closeness between mesenchyma and trophoblast produces definite changes within the syncytioplasma. It becomes brighter and contains less reR whilst free ribosomes are encountered even more frequently. A brighter and rather enlarged matrix is characteristic of the mitochondria. For the first time a tendency of reR lamellae to localize around the mitochondria is observed. Parallel differentiation of the histochemical picture also takes place. In the labyrinth it is already possible to distinguish as peripheral and a central part. AlkP (Kaufmann, 1974) and LAPase (Vollrath, 1965; Kaufmann, 1974) in the central part (present in the rest of the labyrinth also) are strongly manifested (Fig. 19). Now, throughout the labyrinth a well pronounced activity of acP, a5Nase and ATPase (Davidoff, 1970), G6PDH, 6PGDH (Christie, 1968) and HSTDH (Davidoff and Gospodinov, 1971) is also established. Some of the GA are strongly dilated. Numerous polypoid swellings with different content (reR, mitochondria, ribosomes) are observed along the surface of the trophoblast. Cytotrophoblastic cells are encountered very seldom. Glycogen deposits are noted in the cytoplasm of isolated mesenchymal cells.

6. 2. 4. 36th to 50th Day

In general outline by the 36th gestation day, the main guinea-pig placenta terminates its morphological differentiation. Ultrastructural signs of the interlobe and labyrinth, characteristic of the mature placenta, are observed with certainty within 45 days. During this period, the changes involve chiefly the interlobe. Thus many of the secretory granules increase their volume. The Golgi zones give off vesicles deprived of a compact center. Strongly increased are also the lysosome-like bodies, particularly in the marginal syncytium. Microvilli are formed in the labyrinth along the basement surface. Around the 50th day, the guinea-pig placenta displays the picture similar to that established at term.

6.3. Functional Morphology of the Different Regions in the Mature Main Placenta

The fine structure pattern of the mature placenta depends on the fixation method employed (immersion or perfusion), as well as on the osmolarity of the fixing solutions and buffers used. Nevertheless, in the mature placenta it is always possible to differentiate electron-microscopically the following zones: interlobular and marginal syncytium, and labyrinth with a lobar periphery and center of the lobules (Fig. 13a–f). The presence of a middle part of the lobule, uniting the periphery to the center is hardly acceptable.

6.3.1. Interlobium

There is no essential ultrastructural difference between the interlobar and marginal syncytium, and therefore they are considered as an interlobium (see Kaufmann, 1969a; Davidoff and Schiebler, 1970a). These zones are characterized primarily by the absence ob fetal vessels. The syncytioplasma forms broad plaques or thick trabeculae (Fig. 13a, b, 14a). Often, in the plaques numerous nuclei, differing in size, with smooth or uneven surfaces are accumulated. They contain a great quantity of chromatin, unevenly disposed, or massed along the periphery. Arrangement in nuclear rows in the syncytioplasma, beneath the maternal blood spaces is characteristic. The whole interlobe is interspersed with variable sized openings, representing the maternal blood spaces (lacunae). From the syncytial surface enveloping the maternal blood lacunae, numerous microvilli of different length come out. The microvilli on the plasmalemm form numerous invaginations in the syncytioplasma, subsequently detached as bright pinocytotic bubbles. Depending on the manner of killing the animal and the fixation, lipid droplets may be occasionally detected between the microvilli and within the syncytium, underneath the plasmalemm. The peculiarities outlined display a capacity of intense resorption of substrates from the maternal blood (Davies et al., 1961a; Starck, 1975). Apart from the numerous nuclei rich in chromatin, the interlobe syncytioplasma is distinguished by its very strongly developed rough endoplasmic reticulum and abundance of the other cell organelles. The granular endoplasmic cisternae are located mainly around the nuclei. Their membranes are frequently parallel to each other (Fig. 15).

Free ribosomes and polyribosomes are scattered not only on the reR, but also irregularly throughout the entire syncytioplasma. At some points they form fields of different width. This richness in ribosomes and reR is demonstrated light-microscopically with marked basophilia of the capillary-free trophoblast, described by numerous authors (Duval, 1892; Grosser 1909; Wislocki et al., 1946; Amoroso, 1952; Davies et al., 1961b; Starck, 1975; Vollrath, 1965; Kaufmann, 1969a), as well as with the greater RNA quantity established in it (Christie, 1968). Tiny zones filled with GA are detected in the vicinity of the free surface of the syncytioplasmic nuclei. The mitochondria of the latter zone are numerous and unevenly scattered. Occasionally, they form groups. Usually they have an electron-dense matrix and variable size.

The histochemically proved strongly pronounced enzyme activity for the representatives of various metabolite cycles is furthermore related to the considerable amount of reR and mitochondria. SDH (Reale and Pipino, 1959; Christie, 1968; Davidoff, 1970), CO, LDH, G6PDH (Vollrath, 1965; Christie, 1968), IDH, MDH, GDH, αGPDH

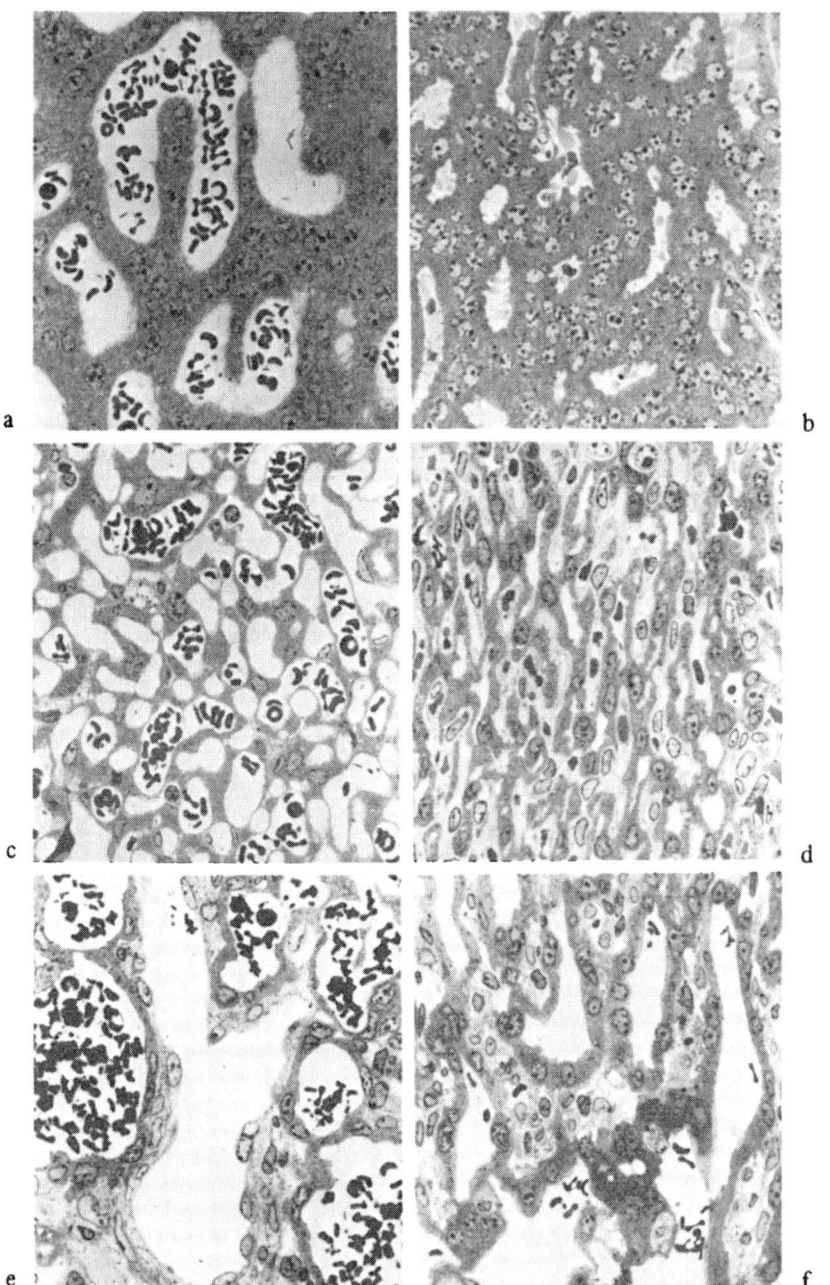

Fig. 13a–f. Comparison of the usual semi-thin section picture of the guinea-pig main placenta after immersion fixation with 6 % glutaraldehyde (right row, b, d, f) with the semi-thin section picture after fetal perfusion fixation with 2.2 % glutaraldehyde (left row, a, c, e). Top row (a, b): interlobium. Middle row (c, d): periphery of the lobe. Bottom row (e, f): centre of the lobe. After immersion fixation (right row), the lumina are largely collapsed, trophoblast and endothelia are swollen. In the lacunae of the interlobium (top right) one sees artificial protrusions of the plasmalemm. After perfusion fixation (left row) the maternal lumina filled with erythrocytes and the rinsed fetal lumina a wide. The separating trophoblastic and endothelial membranes are narrow. This picture is likely to be much more likelife than that after immersion fixation. Semi-thin sections, toluidinblue-pyronin X 370

54

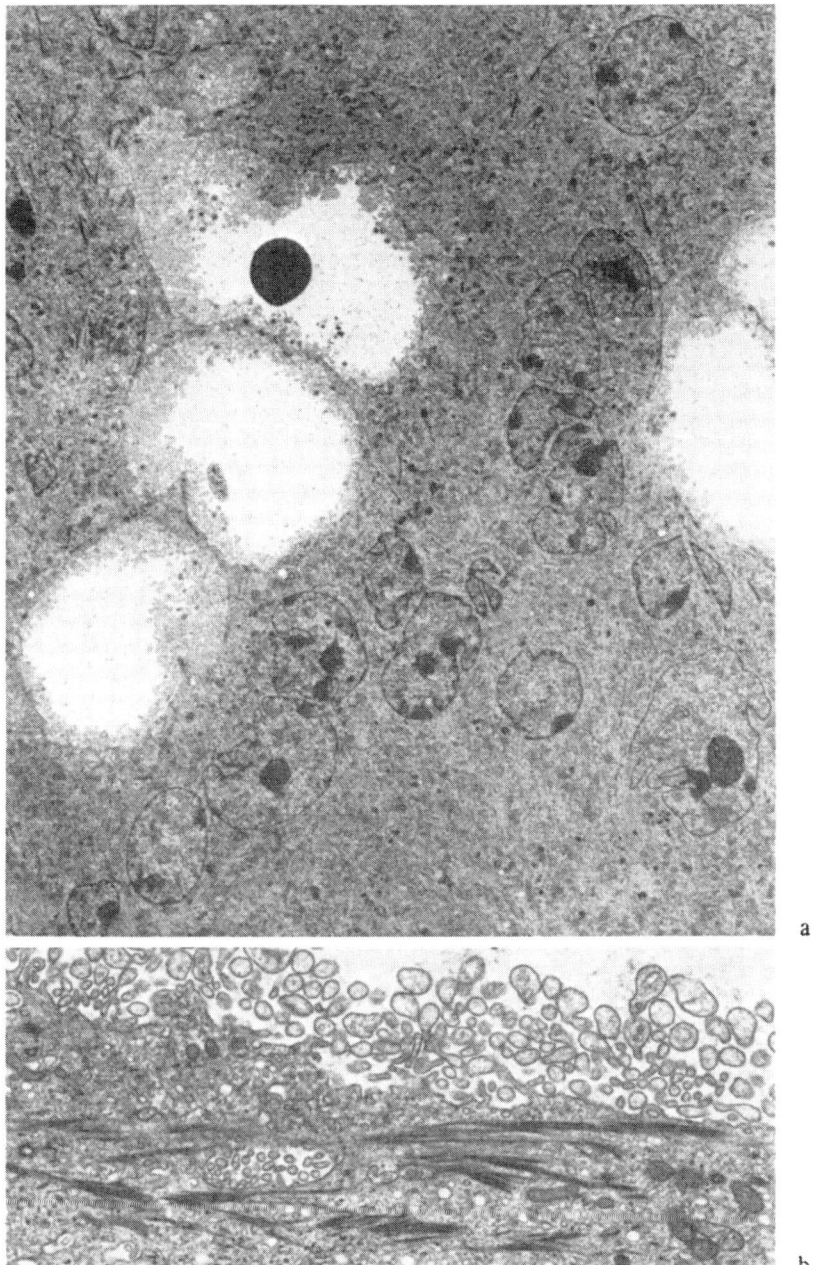

a

b

Fig. 14. (a) Electron-micrographical survey of the interlobium. Within the syncytotrophoblast rich in nuclei one sees lacunae lined by microvilli partly separated by thin septae of syncytoplasm. These are probably branching points of the lacunae. Fixation: fetal perfusion with 2.2 % glutaraldehyde, OsO_4-postfixation. X 1900. (b) Detail electron-micrograph of the interlobium. Under the lacunal surface bearing microvilli one finds a dense network of intracellular filaments. Judging by their location and arrangement, they probably fulfil a mechanical function. Fixation: fetal perfusion with 2.2 % glutaraldehyde, OsO_4-postfixation. X 9700

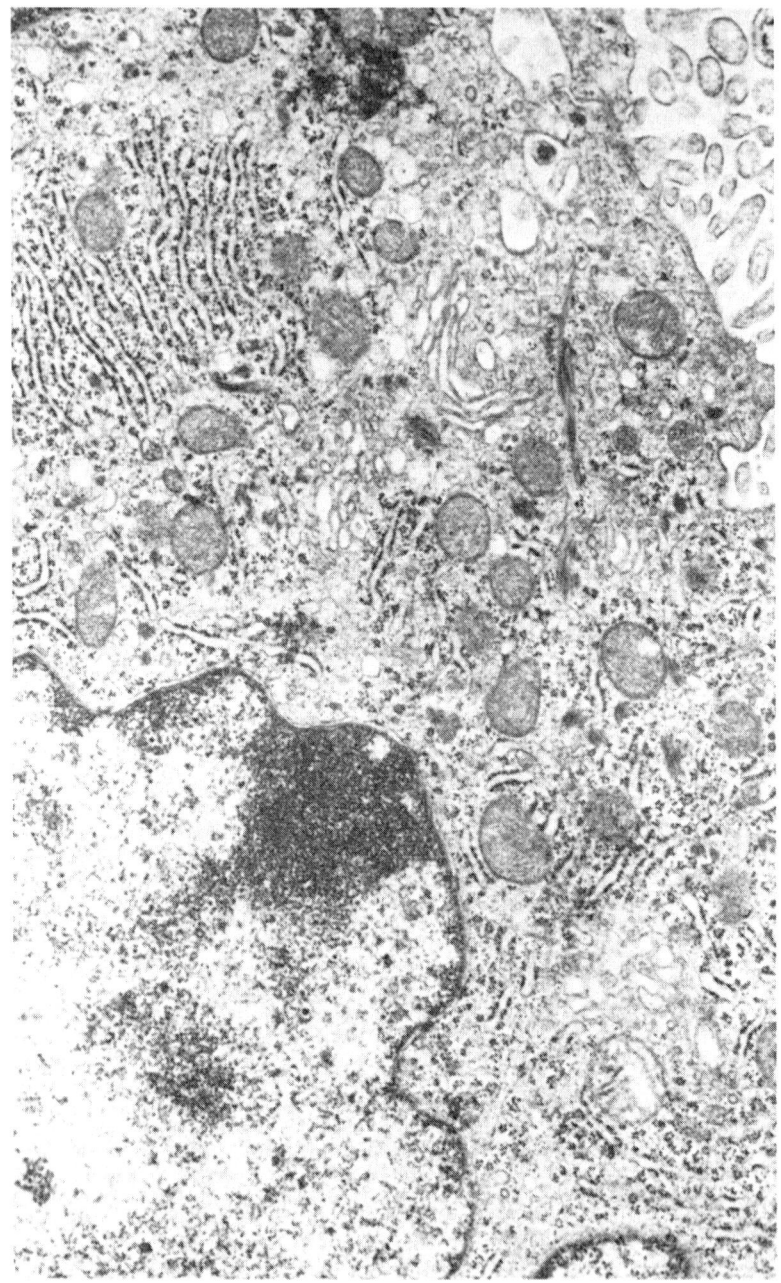

Fig. 15. Detail electron-micrograph of the syncytotrophoblast of the interlobium. Top right the lacunae lined by a border of microvilli. Bottom left a nucleus. In vicinity of the nuclei one always finds large areas of rough e. r. as well as mitochondria. The rough e. r. shows a strinkingly parallel arrangement of its narrow cisternae when well fixed and advantageously cut. The mitochondria have a few cristae only and a dense matrix. Fetal perfusion fixation with 2.2 % glutaraldehyde, OsO_4-postfixation. X 18600

(Vollrath, 1965; Christie, 1968; Davidoff, 1970) have been demonstrated in the inter-lobe (Fig. 19).

The above results show unequivocally that the interlobe is a basic zone in the placenta where aerobe and anaerobe glycolysis (the Embden-Meyerhof pathway and the citrate cycle) are accomplished. Hence, the interlobe appears to be the basic energy center of the main guinea-pig placenta. It is obvious that the great metabolic potential of the interlobe is connected with the intake and processing of substances from the maternal blood, and to the synthesis of structural proteins which are supposed to participate in the build-up of the labyrinth up to the end of term (Kaufmann, 1969a; Davidoff and Schiebler, 1970a).

The numerous GA, adequately developed and having a typical ultrastructure also point to the intense metabolism of the interlobe. Some of them have dilated cisterns, giving off bright vesicles ($0.15-0.25 \mu$) in the direction of the free trophoblast surface, whereas others are unevenly scattered all over the syncytioplasma. Evidently, in this case it is a matter of some type of secretory GA activity, essentially differing from that observed in the period between the 14th and 30th day of placental development. The presence of exclusively strong diversity of vacuolar lysosome-like structures is interpreted as evidence of the great metabolic action of the interlobe. Here the vacuolar apparatus (De Duve and Wattiaux, 1966) is represented by the basic structures listed below, enveloped mainly by a single membrane:

1. Large ($0.3-1.2 \mu$) vesicles with brighter or darker finegranular matrix. Within the vesicles, in addition, a dense central part, multilamellar structures, and a combination between the dense core and the lamellar structure could be also found. The vesicles with osmiophilic thickening may contain fibrils and single lipid drops.

2. Multivesicular bodies ($0.1-0.25 \mu$) which may have an osmiophilic electron-dense central part, myelin figures, and granular material varying in size and electron density.

3. Pinocytitic vesicles of varying size ($0.1-0.3 \mu$) which seem empty.

4. Coated vesicles of different size, most frequently situated subplasmalemmally, and in the vicinity of GA.

5. Small vesicles ($0.1-0.15 \mu$) with a compact center which may be eccentrically positioned, and which resemble dense core vesicles.

6. Autophagic vacuoles, containing identifiable syncytioplasma components (mitochondria, ribosomes, membranes, vesicles, tubules). These structures are of different sizes, and sometimes may be enveloped in a double membrane.

7. Typical, minute dense bodies (0.25μ) are met with rather rarely, and represent a very small part of the lysosomelike structures. Presumably, the latter fact substantiates the weaker acP– and a5Nase-activity, established by Davidoff (1970) and Davidoff and Schiebler (1970a), as compared to the labyrinth. Electron-microscopically (Davidoff and Schiebler, 1970a) the acP-reaction product in the interlobe is deposited only in single Golgi-lamellae and Golgi-vesicles, as well as in part of the lysosome-like structures. The strongest deposition of reaction products is usually observed in the typical dense bodies. It is obvious that for the most part the lysosome bodies, showing deposition of reaction product in separate zones of their matrix, represent heterophagosomes and autolysosomes. In a rather considerable part, it is a matter also of residual bodies, formed during the period characterized by intense secretory processes in the interlobe (14th to 30th gestation day), most probably regulated through autophagic mechanisms (see Smith and Farquhar, 1966). It is of interest to note that in certain zones of the syncytioplasma, most frequently in the marginal syncytium, areas

practically entirely filled with a great diversity of lysosome-like structures are encountered. The impression is that these zones represent stores of primarily residual bodies which do not lend themselves to excretion by the syncytioplasma. It seems that while the interlobe is endowed with an adequate transport capacity, manifested by the strongly pronounced nucleoside phosphatase activity (Christie, 1968) and ATPase (Christie, 1968; Davidoff, 1970; Davidoff and Schiebler, 1970a), its capacity to excrete substances included in the lysosome-like structures is heavily reduced. Most likely, the latter fact is corroborated, for other substances too, by the presence of syncytioplasmic protrusions (Kaufmann, 1969b; Davidoff and Schiebler, 1970a) along the surface surrounding the maternal lacunae. These protrusions may contain various cell organelles. Kaufmann (1973b) and Kaufmann et al. (1974) provoked experimentally a manifold increase in syncytioplasmic protrusions after blocking the Embden-Meyerhof pathway with monoiodine acetate and sodium fluoride.

The mature placenta, and more particularly the interlobe have strong AChE activity (Goutier-Pirotte and Gerebtzoff, 1955; Gerebtzoff, 1957; Perrotta and Lewis, 1958; Davidoff, 1970). Davidoff (1970) and Davidoff and Schiebler (1970a) proved at electron microscope level a diffuse reaction product deposition in the syncytioplasma. A clearcut deposition is detected along the marginal membrane and mitochondrial cristae. The above cited authors are of the opinion that in the latter case it is a matter not of a real AChE activity, but rather of manifestation of the activity of some of the ferro-ferri-redox systems of cytochromes from the biological oxidation respiratory chain.

The mature placenta interlobe is traversed by osmiophilic filamentous bundles usually running immediately under the surface of the lacunae (Fig. 14b). Nearing term, the quantity of filaments augments. Their disposition beneath the free surface of the trophoblast near the maternal lacunae, is characteristic. Functionally the ultrastructural peculiarity referred to may be assigned to the fibrous framework of the trophoblast, which has primarily supporting function.

6. 3. 2. Transitional Zone

The transitional zone between interlobe and labyrinth is identified electron-microscopically by the presence of few fetal capillaries. The syncytioplasma of the transitional zone shows the peculiarities of the two zones it unites. A characteristic feature is the gradual thinning of syncytioplasmic trabeculae towards the periphery of the labyrinth. Its electron density decreases within the trabeculae. In some of the latter all the organelles are strongly reduced, whereas in others the reduction is less pronounced. The number of cellular nuclei likewise diminishes. The microvilli in the maternal blood lacunae become shorter and are spaced at a greater distance from each other. Davidoff and Schiebler (1970a) explain the appearance of bright cells, disposed along the outer surface of the basement membrane of the vessels, as a characteristic feature of this particular zone. These cells have desmosome contacts with the trophoblast. Hitherto, their nature has not been clarified. Wandering decidual cells (Schiebler and Knoop, 1959), allantoic cells (Franke, 1969), mesenchymal cells or especially cytotrophoblast cell residues come into consideration.

As underlined by Kaufmann (1969a), the transitional zone represents the place where the interlobe is converted into a labyrinth (the labyrinth increases at the expense of the interlobe). In addition to being morphologically differentiated, this zone shows a number of histochemically-proved metabolic peculiarities (Fig. 19). Here G6PDH and hexokinase show marked activity. In addition, a well pronounced acP activity is also recorded (Davidoff, 1970), while we succeeded in demonstrating very strong alkP activity only in unfixed cryostat sections. The studies just mentioned show that the transitional zone is characterized by intense carbohydrate metabolism. Here the marked alkaline and acid phosphatase activity should be attributed rather to the present adequately developed fetal vessels. As a matter of fact, the fetal arteries are situated at the junction between interlobe and labyrinth (Kaufmann, 1969a, 1973a, 1975b). They give off arterioles which, following dichotomic division, are transformed into capillaries (Fig. 8b). The endothelial cells of the arterioles are voluminous ($5-10\ \mu$ high in the nuclear zone, and $1-2\ \mu$ in the more distant areas. Unlike the arteries from which they originate, the long, flat endothelial outgrowths of the arterioles are disposed in parallel to the longitudinal axis of the vessel. That is why on the cross-section of an arteriole it is possible to see 3 to 5 times thicker endothelial fragments (some of them containing nuclei), and numerous cytoplasmic outgrowths from endothelial cells. The supranucleary situated cytoplasm and organelles are edematously altered. This could eventually be related to blood-flow regulation by the endothelial cells. The cytoplasm of the other endothelial cells is electron-dense and contains numerous organelles (small dense mitochondria, strongly pronounced basally situated reR, multiple tiny GA, pinocytotic vesicles and polyribosomes). Smooth muscular cells and single adventitial connective-tissue cells are established beneath the basement membrane.

The precapillaries' cytoplasm in the transitional zone is brighter and rarely contains reR, mitochondria and GA. In some places the quantity of vesiculous or tubuluar smooth endoplasmic reticulum is greater. A characteristic sign is presented by single smooth muscle fragments, situated along the outer surface of the basement membrane. Sometimes they may be enveloped by the latter.

The transitional zone vessels are often positioned in thin, $10-20\ \mu$, bandlike connective-tissue septa orientated tangentially to the lobular surface. The latter are made up of a connective tissue network and free connective-tissue cells situated in its openings (see connective tissue of the lobe center).

6. 3. 3. Periphery of the Lobe

The labyrinth (capillarized syncytium) of the mature placenta (see Enders, 1965; Kaufmann, 1969a; Davidoff and Schiebler, 1970a) may be divided into peripheral, middle and central parts (Fig. 8a, b). The presence of thin syncytioplasmic trabeculae, which on one side come in contact with varying sized maternal blood spaces (free surface of the trophoblast), and on the other with the fetal vessels (basement surface of the syncytioplasma), is characteristic of the peripheral labyrinth. (Fig. 16a, b). The basement membranes of the vessels and trophoblast merge at a distance of varying length.

Basally, the trophoblast forms microvilli-filled invaginations, oriented towards the vessels (Müller and Fischer, 1968). The microvilli on the free surface are shorter and more rarely spaced apart than in the interlobe. The rather weakly developed reR, often located around the mitochondria accumulate near the nuclei, is characteristic of the syncytioplasma. The mitochondria are larger, and with well outlined cristae. In the zones free of mitochondria and endoplasmic reticulum, occasionally ribosomes are

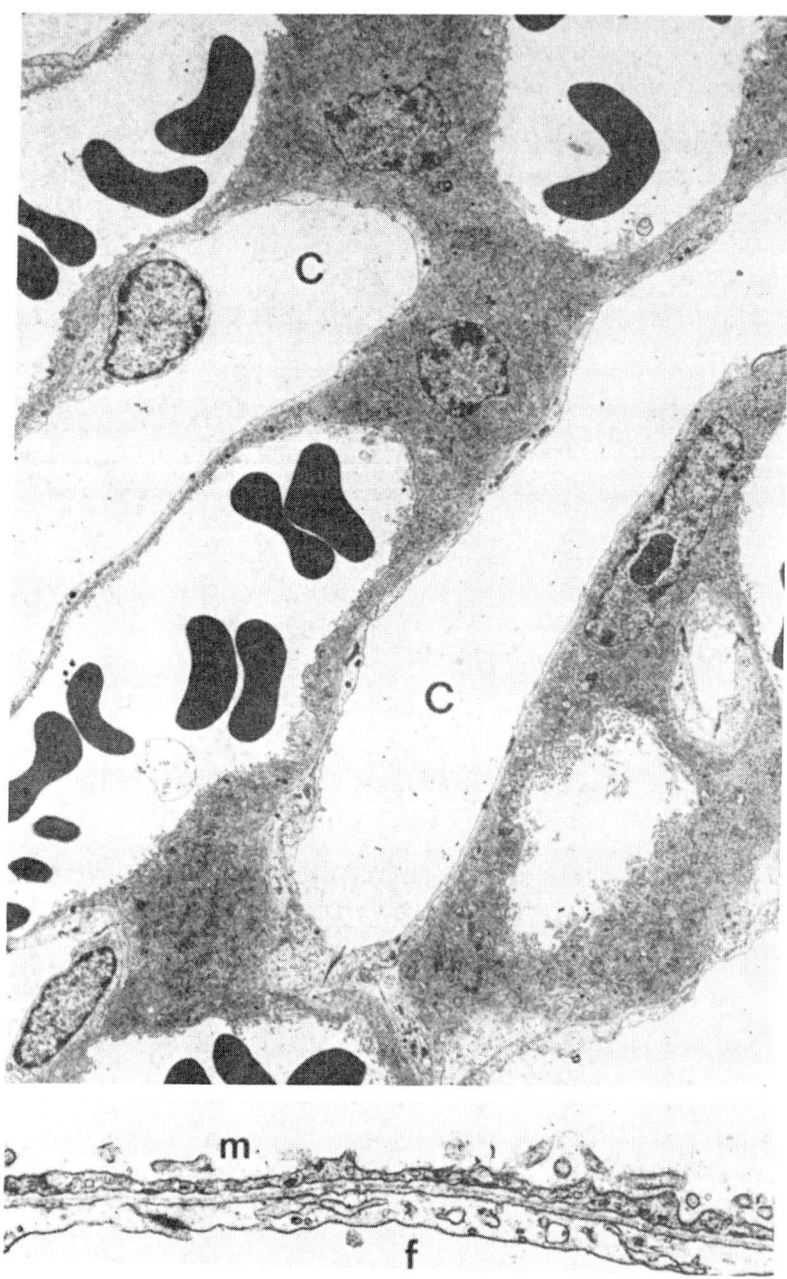

Fig. 16a and b. Electron-microscopical representation of the periphery of the lobe. (a) survey. The capillaries (C) free of erythrocytes are lined by an extremely thin endothelium. The syncytium encompassing the maternal lacunae containing erythrocytes is also very thin over considerable lengths but thicker in the gussets. Such an image with large, extremely thin syncytoendothelial membranes is only obtainable after approximately isotone perfusion-fixation. With immersion-fixation these lumina collapse (cp. Fig. 17). (b) section from the survey above. The syncytial lamella, bearing a few microvilli and nearly free of organelles (above), the slightly thicker endothelium and the fused basal membranes of endothelium and syncytium add up to a layer of $0.48\,\mu$ only separating maternal (m) from feta blood. X 18000. Fixation (14a and b): fetal perfusion with 2.2 % glutaraldehyde, OsO_4-postfixation

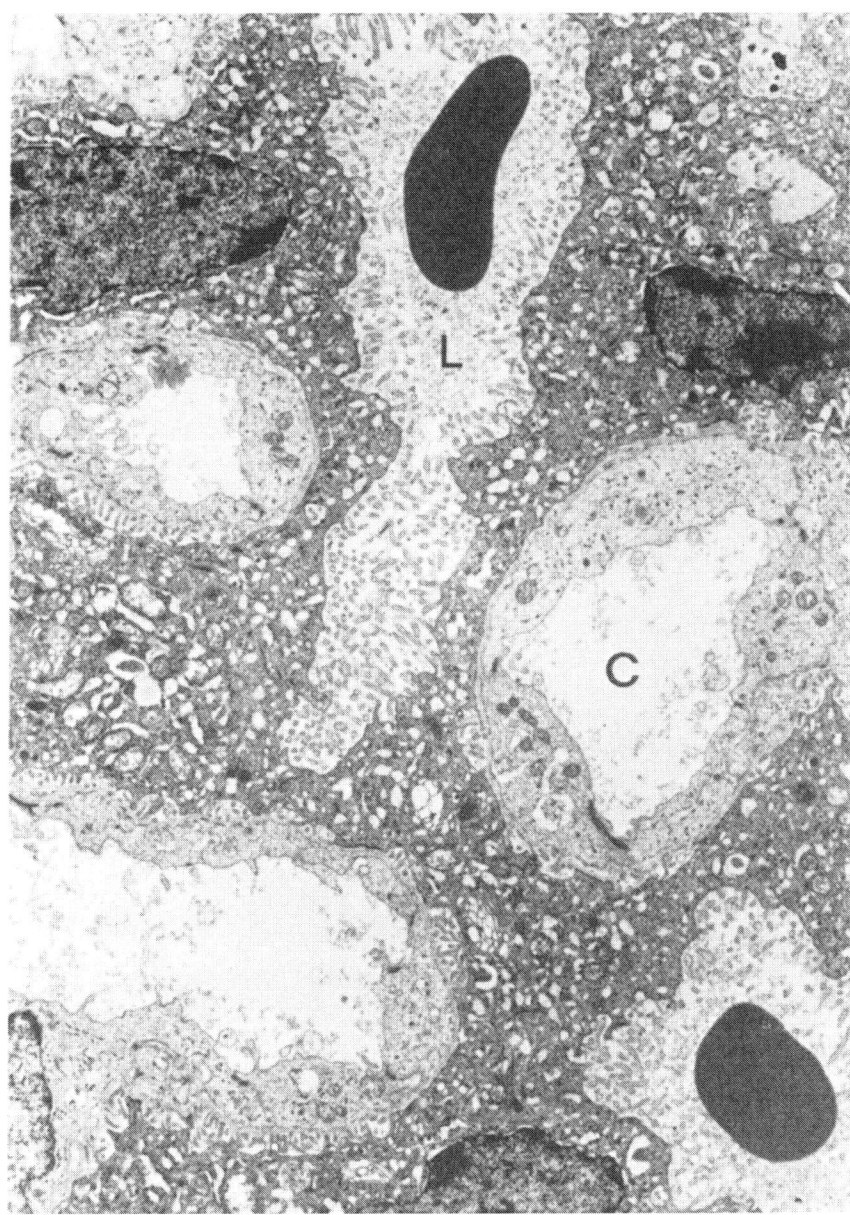

Fig. 17. Electron-microscopical survey of the periphery of the lobe after immersion fixation with 6 % glutaraldehyde and OsO_4-postfixation. Compared with the preceding picture of a placenta fixed by perfusion (Fig. 16), the capillary lumina (C) and the largely collapsed lacunae (L) are significantly smaller here. The capillary endothelium is diffusely swollen. In the trophoblast, dilatation of all membraned organelles is conspicuous. The diffusion distance from the maternal to the fetal blood is even in the thinnest spots manifold longer than after perfusion. X 5700

aggregated. Golgi zones are met with often. They are arranged either perinuclearly, or chiefly in the basal zone of the trophoblast (see Enders, 1965). GA give off bright vesicles, probably connected with the transport of substances in the syncytioplasma.

The finding of numerous dense bodies is characteristic of the labyrinth syncytioplasma. Most likely, their presence conditions the stronger acP- and a5Nase activity of the labyrinth (Davidoff, 1970). Autophagic vacuoles are rarely encountered. It is interesting to note that the Golgi zones of the labyrinth do not show acP activity (Davidoff and Schiebler, 1970a), whereas the lysosomes and Golgi zones of the endothelial cells of the vessels display an abundant deposition of reaction product. The same authors described reaction deposition for ATPase and thiamine-pyrophosphatase along the basement membranes, more particularly in the places where the syncytioplasma forms basal microvilli. Obviously, in the zone under review transport processes of considerable intensity develop. This is supported by the fact that alkP activity is established exclusively in the capillarized sections of the trophoblast (Hard, 1946; Wislocki and Dempsey, 1949a, b, c; Davies et al., 1961b; Christie, 1967; Davidoff, 1970), as well as by the great diversity of pinocytotic vesicles.

In the peripheral labyrinth the very thin syncytioplasmic trabeculae are particularly impressive (Fig. 16b); they are situated·over delicate outgrowths from the endothelial cells of fetal capillaries, and resemble very strongly the human placenta epithelial plates (Schiebler and Kaufmann, 1969). It is by no means excluded that the gas exchange between mother and fetus takes place exactly in these zones. Apparently, the periphery of the labyrinth plays an important rôle in the glucose metabolism, indicated also by the histochemically demonstrated G6PDH and 6PGDH (Vollrath, 1965; Christie, 1967; 1968; Davidoff, 1970), SDH, LDH, GDH, βHBDH (Vollrath, 1965; Davidoff, 1970) (Fig. 19), and in the steroid metabolism as well, which is taken up by the labyrinth after its formation (Vollrath, 1965; Davidoff and Gospodinov, 1971). Evidently, the substances supplied to the placenta from the fetus have an essential bearing on steroid metabolism in the mature placenta (see Davidoff and Gospodinov, 1971).

The characteristic fine structural features of the labyrinth vessels endothelial cells also point to the presence of active transport processes between maternal and fetal blood. In the periphery of the lobule, the precapillaries are deprived of their muscle cells, and assume the characteristics of capillaries. Centrally, the endothelial cells account for an increased amount of reR, with ensuing denser cytoplasm. Golgi zones and smooth eR are encountered more rarely, while the pinocytotic vesicles are more numerous.

6. 3. 4. Middle of the Lobe

The middle part of the labyrinth is characterized by its more uniform structure. Similar sized openings are formed in the maternal blood spaces and in the fetal vessels. Both free and basal surfaces of the syncytoplasm possess a reduced quantity of microvilli. A tendency for a slight reR increase is noted which is more strongly pronounced in the direction of the lobe center. Histochemically, the middle part of the lobe reveals no essential differences in comparison with the lobar periphery. Here again, the tendency for alteration of organelles in the endothelial cells of the capillaries persists, and in the vicinity of the lobe center the onset of a gradual increase in the number of mitochondria is observed.

6. 3. 5. Centre of the Lobe

The central labyrinth is characterized by a gradual syncytial trabeculae thickening, and distension of maternal and fetal blood spaces (Fig. 13e, f, 18). The free surface of the trophoblast contains a few microvilli. RER increases in the syncytoplasm which, depending on the fixation may be either strongly dilated and vacoulized, or, in case of a more successful fixation, composed of more compactly disposed short membranes covered by numerous ribosomes, is characteristic (Fig. 18). Often, areas are encountered wherein accumulation simultaneously of mitochondria and endoplasmic reticulum is observed. The number of free ribosomes and polyribosomes is substantially augmented. Also fields with smooth endoplasmic reticulum occur. The Golgi zones are very well developed, but presumably they have weaker activity compared to the peripheral labyrinth. Basal microvilli are more frequently encountered. In addition, in the syncytioplasma there are rather small dense lysosomes, many pinocyte vesicles (mainly coated, and smooth as well), and numerous other, diffusely scattered smooth vesicles. Single free lipid drops are also found. Histochemically, the central labyrinth, apart from the peculiarities mentioned insofar as periphery is concerned, exhibits marked SDH, GDH (Davidoff, 1970), LAPase (Vollrath, 1965; Kaufmann, 1974) (Fig. 16), which outlines this particular zone as the site of intense protein and nucleic acide metabolism, re-synthesis of proteins inclusively. The fact that the greatest amount of lipids and βHBDH were found here (Wislocki et al., 1946; Davies et al., 1961b; Vollrath, 1965; Davidoff, 1970) underscores the rôle played by the lobe center in lipid metabolism.

The appearance of pericytes in the stage of transition to the lobe center is the basic characteristic feature of the capillaries which, at this point, should be coined as postcapillaries or venous capillaries (Fig. 8b). Here the endothelial cells display the strongest electron density of the cytoplasm, and have adequately developed reR and numerous mitochondria. Golgi apparatus and smooth endoplasmic reticulum are recorded very rarely. Towards the center, these capillaries merge into venules, augmenting their lumen and number of pericytes, with ensuing appearance of isolated smooth muscle cells. In the same direction, the cytoplasm of the endothelial cells becomes brighter. The granular endoplasmic reticulum increases to maximum quantities, while the number of Golgi apparatus and the smooth endoplasmic reticulum amount — to the extent found in none of the other parts of the vascular system hitherto described (Fig. 18) This exceptional abundance of organelles in the vessels of the lobe center complies with the physiological data concerning the rich oxygen supply of this particular zone.

The ultrastructural results and histochemical data referred to above prove the participation of the lobe center in the synthesis processes related to the materno-fetal protein exchange, and in the realization of maximum transport. The fact that oxygen pressure and concentration of carbohydrates and amino acids increase from the arterial towards the venous side, whereas partial carbon dioxide pressure and concentration of split products from amino acids and nucleic acids decrease in the same sense, is an important characteristic feature of the guinea-pig capillaries.

A greater connective tissue accumulation is also established in the lobe center. Connective tissues are usually located around the fetal veins, venules and postcapillaries. In the same area, single connective-tissue proliferations may contribute to the formation of structures resembling placental villi. The connective tissue in this particular zone, similarly to that in the vicinity of the major blood vessels and in the transitional

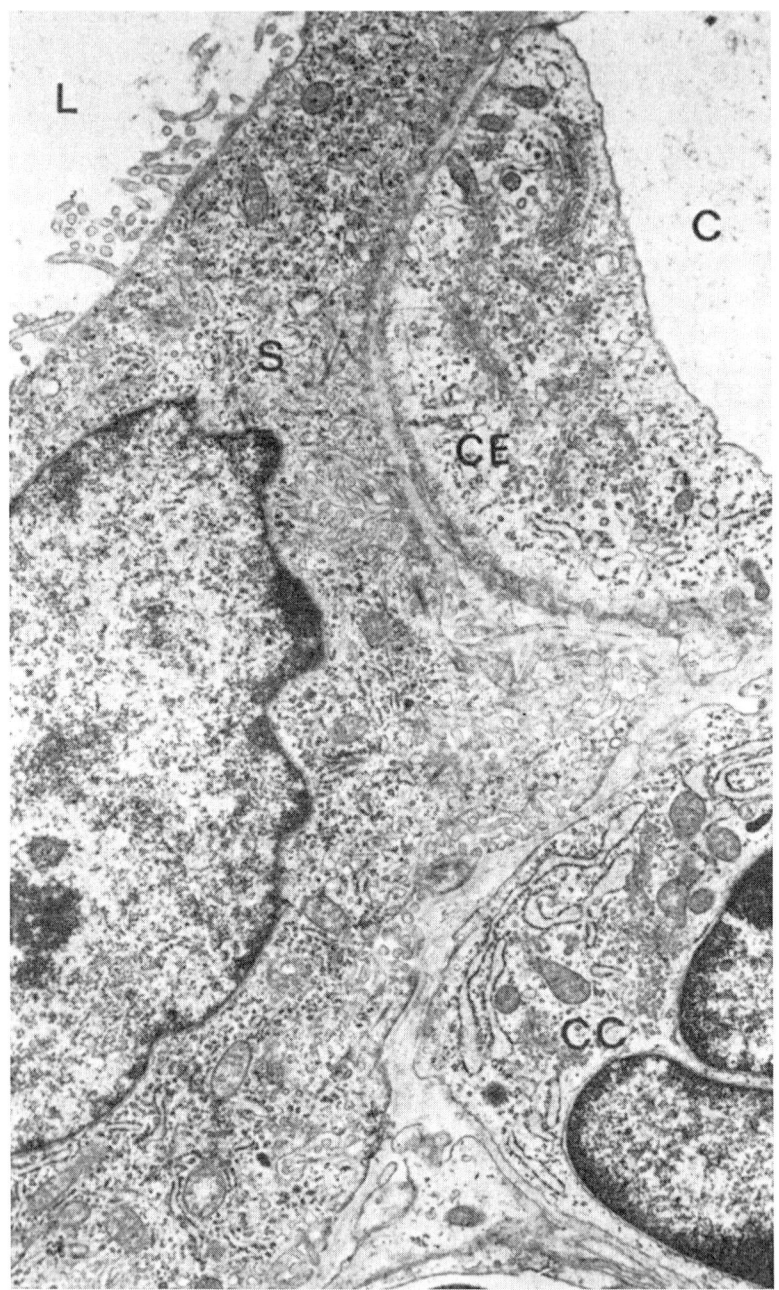

Fig. 18. Electron-micrograph of the centre of the lobe. The distance between fetal blood (*C*) and maternal blood (*L*) is considerably larger than in the periphery of the lobe, since both capillary endothelium and syncytotrophoblast are thicker. The capillary endothelium (*CE*) shows more cell organelles than in the periphery. Also in the syncytotrophoblast (*S*) the rough e. r. is more abundant than in the periphery of the lobe. Bottom right one recognizes part of a connective tissue cell (*CC*). Perfusion fixation with 2.2 % glutaraldehyde, OsO$_4$-postfixation. X 10700

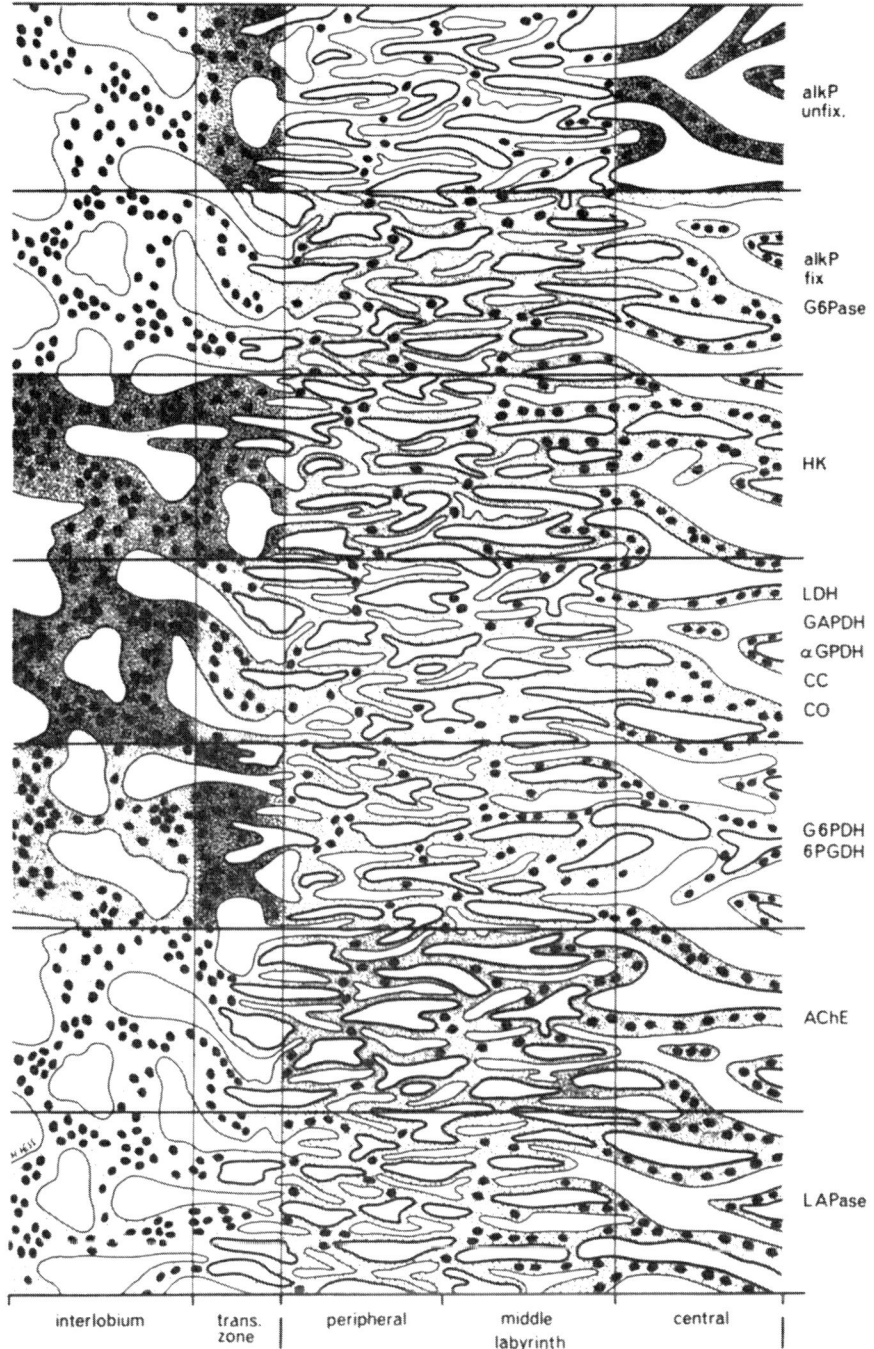

Fig. 19. Strongly schematized representation of the activity of several enzymes in the interlobium, transitional zone, periphery, middle and centre of the lobe. The enzymal activity is symbolized by the density of the dot-screen. *alkP unfix.*: alkaline phosphatase in the unifixed cryostat section. *alkP fix.*: alkaline phosphatase in the fixed cryostat section

zone, is made up of a connective-tissue network and free connective tissue cells, situated within the latter.

The cells of the connective-tissue network are endowed with an abundant cytoplasm. It gives off numerous outgrowths which unite with the outgrowths of the neighbouring cells. Thus narrow spaces are formed. In the composition of the above connective tissue network it is possible to distinguish: 1) phagocytizing reticular cells (characterized by electron dense mitochondria, single endoplasmic reticulum lamellae, small Golgi zones, many vacuoles chiefly at the outgrowth ends, varying sized lysosomes and numerous pinocytotic vesicles beneath the plasmalemma); 2) small reticular cells (rarely met with, and possessing a great quantity of vesiculous granular endoplasmic reticulum); 3) typical fibroblasts (single, spindle-shaped with few outgrowths and a large amount of granular reticulum, distended in some places) (Fig. 18). Usually, they are situated in the vicinity of large vessels where a greater quantity of collagen is also established; 4) mesenchymal cells (small, poor in organelles with filiform outgrowths); they are practivally absent in the secont half of pregnancy.

The free connective-tissue cells show a much greater morphological variety. Among them histiocyte-like cells, macrophages, plasma cells, mastocytes, and lymphocytes are observed. Histiocyte- and macrophage-like cells hardly lend themselves to differentiation. Between these two basic forms histiocytes (vacuolized cytoplasm, single cisternae of granular endoplasmic reticulum and strongly developed smooth endoplasmic reticulum) and phagocytes (heavily granulated with numerous secondary lysosomes), a number of intermediate forms occur showing that histiocytes and phagocytes may be assigned, in principle, to the different functional stages of cells of one and the same type.

According to Enders and King (1970) and King and Enders (1970c, 1971) the role played by these phagocytizing cells consists in taking up the maternal proteins which have passed through the trophoblast without undergoing splitting. Their dominant positioning in the center of the lobe which proves to be, histochemically and electron-microscopically, the basic zone of materno-fetal protein and amino acid transport seems logical on first sight. However, it is difficult to admit that these cells play a decisive role by forming a barrier to the transfer of maternal proteins in the fetus. The higher number of phaging reticular cells here seems practically insufficient since the greater part of the materno-fetal exchange surface consists of syncytio-capillary contacts wherein fetal connective tissue cells are not included.

6.4. Subplacenta and Junctional Zone

6. 4. 1. Development and Regression of the Subplacenta

The development of the subplacenta starts on the $15^{th}-16^{th}$ day of gestation when the top of the central excavation broadens in a fungoid way while its part lying towards the fetus (connected with the allantois) is narrowed by lateral trophoblastic growths and forms what is referred to as central mesenchymal axis (Fig. 3f–h). The layer of trophoblast now separating the fungoid-shaped part of the central excavation basally from the junctional zone is termed the roof of the central excavation according to Duval (1892). It consists of a cytotrophoblast layer where it borders the mesoderm and of a layer of syncytium towards the junctional zone. From the latter, basal sprouts of syncytium invade the junctional zone and the decidua (Fig. 4).

Connective tissue invades the developing main placenta laterally from the central mesenchymal axis on the 16th day. At the same time, septa of connective tissue originating from the fungoid-shaped region of the central excavation also invade the roof of the central excavation which is thereby deformed and becomes starshaped (Creighton, 1878; Ercolani, 1879; Bischoff, 1866; Laulanié, 1886). The two-layered trophoblast stratum which is divided into separate lobes by these septa of connective tissue is designated subplacenta. This terminus was introduced by Minot (1889) for a similarly located but structurally differing formation in the rabbit. Later on it was applied to the guinea-pig.

As the growth of the main placenta exceeds that of the subplacenta by far during the following days, the attachment area of the placenta on the uterus wall does not increase to the same extent as the size of the placenta. The placenta acquires a stem. The stem of the placenta consists of the subplacenta with the enlarged part of the central excavation centrally covered by a thin layer of junctional tissue through which the maternal vessels run into the main placenta, bypassing the subplacenta. The forming of the placental stem is already clearly distinguishable on the 19th day and continues till the end of pregnancy.

The increase in size of the subplacenta is due to (1) the proliferation of the cytotrophoblast, (2) its transformation into syncytium, (3) the continuing invasion of ramified mesenchymal septa out of the central excavation into the trophoblast. Thereby the subplacenta is increasingly split into lobes. On the 46th day they have a largely constant diameter of 200 to 250 μ, but length and number keep increasing owing to proliferation of trophoblast and the growth of new septa. After the 46th day they decrease in size due to degenerative processes. Each lobe consists centrally of syncytium which is perforated by a net of more or less developed lacunae. Each lobe is wholly covered by a layer of cytotrophoblast whose thickness may differ, however. On the outside there are the septa of connective tissue which will be invaded by fetal vessels later on (see Davies et al., 1961a, b).

The increase in size of the subplacenta happens in phases (Uhlendorf and Kaufmann, 1976). The first phase lasts until the 18th day resulting in a broadening of the top of the central excavation and an extension of the subplacenta without substantial increase in thickness. Its volume remains constant up to the 22nd day after which it grows in width, slowly at first (Table 12).

From the 26th day onwards it gets thicker through longitudinal growth and further ramification of the lobes. This second phase, lasting up to the 41st day, sees the most decisive increase in volume of this organ, up to the 40-fold initial volume on the 16th day. The size of the central excavation remains roughly constant throughout this phase. On the whole, the direction of growth is more lateral than basal so that its originally bowl-shaped form flattens more and more. As from the 46th day in the third phase the volume of the subplacenta decreases again owing to degenerative processes.

Concomitant with this development, there are characteristic alterations in terms of mass-proportions of cytotrophoblast and syncytium (Uhlendorf and Kaufmann, 1976): up to the 26th day, the cytotrophoblast is bigger than the syncytium (Table 12). After this the amount of syncytium (including the lacunae) increases rapidly due to diminishing proliferation of the cytotrophoblast and a higher syncytial fusion-rate so that 80 % of the subplacenta's volume is taken by syncytium around the 40th day. Af-

Table 12. Statistically unverified data on development of the subplacenta. The data in the first nine columns are based on Bouin-fixed samples embedded in paraffin. The unavoidable artificial shrinking relativates the absolute meaning of the data. The data on secretory granules and glycogen (last two columns) were won by means of electron microscopy from glutaraldehyde-fixed samples. Owing to degenerative phenomena, no absolute length and volume data can be made from the 52nd day onwards. As from the 60th day there are no structures whatsoever which can be evaluated by means of morphometry

Age in days	Diameter (mm)	Thickness (mm)	Volume (mm³)	Cytotrophoblast (%)	Syncytium incl. lacunae (%)	Connective tissue incl. fetal vessels (%)	Cytotrophoblast (mm³)	Syncytium incl. lacunae (mm³)	Connective tissue incl. fetal vessels (mm³)	Share of secretory granules in syncytioplasma (%)	Share of glycogen in syncytioplasma (%)
16	2.5	1.0	2								
18	3.5	1.0	2.5	50	40	10	1.25	1.0	0.25		
20	4.0	1.0	4	53	37	10	2.12	1.48	0.4		
22	4.0	1.0	4.5	52	38	10	2.34	1.71	0.4		
24	4.1	1.0	5.5	50	40	10	2.75	2.20	0.5	0.9	9.3
26	4.3	1.0	7	47	43	10	3.29	3.01	0.7	2.2	15.3
28	4.6	1.2	10	34	56	10	3.4	5.6	1.0	2.3	16.7
30	5.2	2.1	32	25	65	10	8.0	20.8	3.2		27.6
32	5.8	2.8	50	23	67	10	11.5	33.5	5.0		
34	6.6	2.8	60	21	69	10	12.6	41.4	6.0	3.5	5.1
36	7.7	2.8	68	20	70	10	13.6	47.6	6.8	3.8	8.7
38	8.3	2.8	74	19	71	10	14.1	52.5	7.4	3.8	11.2
40	8.4	2.8	82	18	72	10	14.8	59.0	8.2	5.6	13.7
42	8.3	2.8	83	18	72	10	14.9	59.8	8.3		
44	8.2	2.8	83	18	72	10	14.9	59.8	8.3		
46	7.9	2.7	80	18	72	10	14.4	57.6	8.0		
48	7.5	2.4	72	19	71	10	13.7	51.1	7.2	9.4	19.2
50	7.0	2.0	60	21	69	10	12.6	41.4	6.9		
52				24	66	10				4.5	
54				26	64	10				4.5	19.9
56				28	62	10					
58				31	59	10				3.3	15.3

ter the 47th day, degenerative processes taking place mainly in the syncytium lead to a slight approximation of the volume of syncytium and cytotrophoblast. The relative proportions of connective tissue and fetal vessels in the subplacenta remain nearly constant at nearly 10 % throughout development.

If one regards the absolute volume changes of cytotrophoblast and syncytium including the lacunae on realizes that both increase upt to the 45th to 47th day, the syncytial part, however, markedly more than the cytotrophoblastic parts as from the 27th day. After the 47th day, the syncytium suffers a heavy loss in volume whereas the cytotrophoblast regresses only moderately. From the 52nd day onwards, processes of degeneration in syncytium, cytotrophoblast and connective tissue are very pronounced. Therefore statements concerning volume can no longer be made.

Since the subplacental syncytotrophoblast is the part of the whole placenta lying most basally, i. e. nearest to the decidua, its lacunal system is the first to join with the maternal circulation. Up to the 20th day, approximately, this maternal circulation is fully maintained. Later on erythrocytes may still be detected in the lacunae, but the surrounding plasma has coagulated fibrously in most areas of the subplacenta.

Maternal circulation in the subplacenta after the 20th day is therefore doubtful. Around the 28th day is has ceased in all regions of the subplacenta. The erythrocytes have mostly disintegrated, the plasma has turned into fibrin. Both condense increasingly with the shrinking of the lacunae to form an unhomogenous electron-dense mass (Fig. 20c).

Well after the maternal circulation has come to a standstill, new lacunae are still formed regularly (Fig. 20a). Electronmicrographs revealed that the formation of lacunae is constantly preceeded by the deposition of glycogen in large areas which are then surrounded by small vesicles (Wolfer and Kaufmann, 1976). The vesicles merge, forming a double membrane around such glycogen areas (Fig. 20b). The inner membrane decomposes together with the glycogen while the outer membrane persists as the new lacunal wall. Consequently free glycogen particles may frequently be found in the newly built lacunae, which mingle with the precipitated remnants of cytoplasm in the lacunae, leading anew to the deposition of electron-dense material. This tendency to the formation of lacunae differs strongly locally, but a preferential location cannot be stated. But it is striking that this happens in phases. An early lacunae formation phase occurs before the forming of the subplacenta in the roof of the central excavation. Another one is to be observed between the 23rd and 27th day. Later on we find largely regression of the lacunae which shrink with their content condensating. A third phase can regularly be stated between the 38th and 40th day. Comparison with the volume development of the subplacenta shows that phases of forming of lacunae alternate with phases of volume increase. For the time being it cannot be decided whether this is a fortuitous coincidence or whether both processes are related to each other. The important fact, however, is that maternal circulation in the subplacenta comes to a standstill between the 20th and 27th day and never starts again, also not in the newly built lacunae.

In contrast, the fetal vascularization commences much later (Wolfer and Kaufmann, 1976). Although there are a few fetal vessels in the central excavation at the time of the forming of the subplacenta, they do not come into contact with the trophoblast. Between the 23rd and 27th day — the period during which the maternal circulation recedes — the first fetal vessels penetrate deeper between the trophoblast lobes. The fetal vascularization of the subplacenta is only completed on the 32nd day. After the 55th day the vessels are obliterated in the course of the degenerative processes in the subplacenta. From the 58th day onwards there are hardly any more fetal vessels to be detected.

It may therefore be gathered from these findings that the subplacenta is at no time sufficiently vascularized maternally and fetally synchronously (Wolfer and Kaufmann, 1976). Consequently, it cannot be considered an organ for feto-maternal exchange of substances as has been suggested now and then. The vascularization seems to serve only the nutrition of the subplacental tissue.

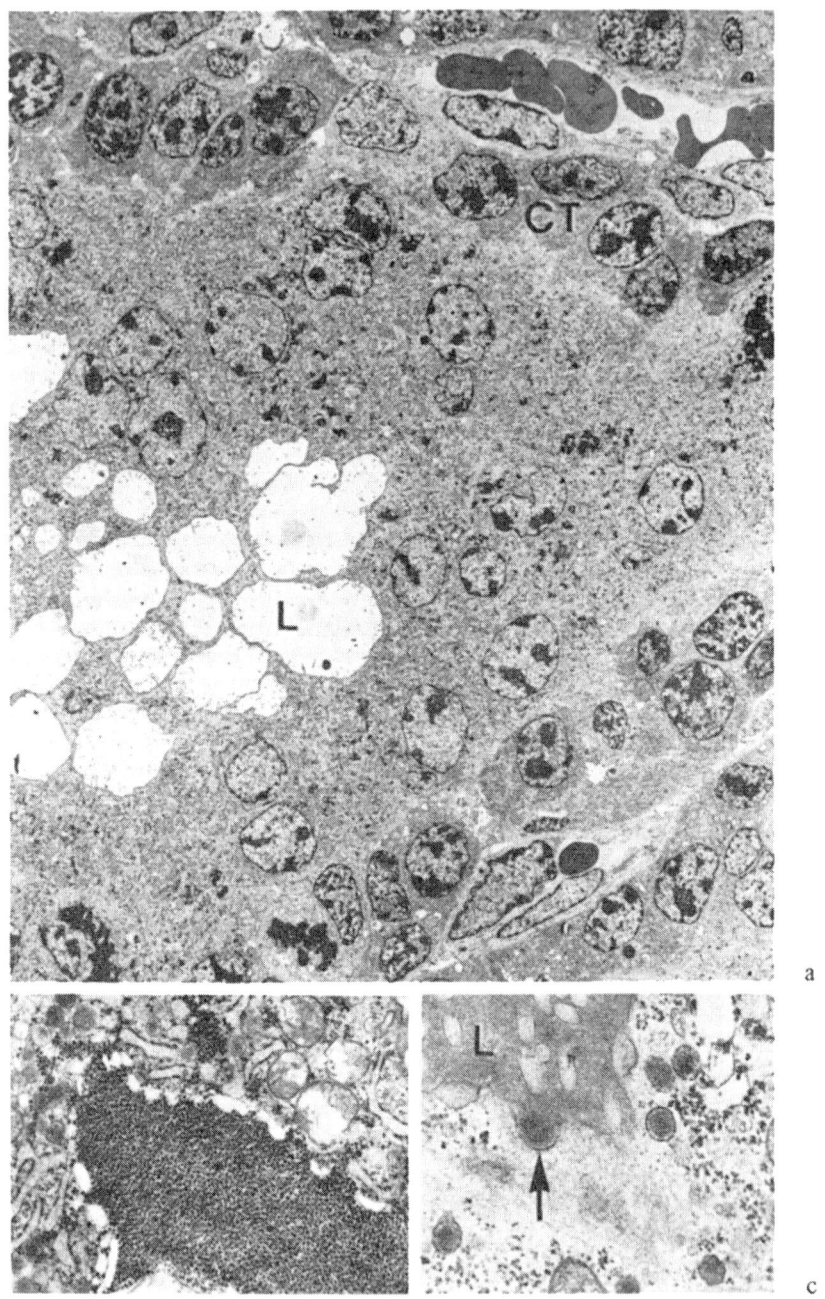

a

b

c

Fig. 20. (a) Electron-microscopical survey from the subplacenta of the guinea-pig from the 33rd
day. Above and on the right one recognizes thin septae of connective tissue with capillaries (C) with
erythrocytes. At this stage the connective tissue is surrounded by a mainly single-layered stratum
of cytotrophoblast (CT). Due to the abundance of ribosomes the cytotrophoblast presents higher
electron-density than the syncytotrophoblast. The latter fills the centre of the picture. On the left
one recognizes several lacunae (L) without circulation within the syncytotrophoblast. X 1500
(b) Detail micrograph from the 38th day. A large and compact glycogen area is separated from the
syncytoplasm by a row of vesicles. After fusion of these vesicles the glycogen is surrounded by a

6. 4. 2. Functional Significance

In the course of early histochemical investigations of the subplacenta (Davies et al., 1961a, b), a striking amount of PAS-positive substances was observed in this organ. Control tests with α-amylase showed that these substances when fine-grained were PAS-positive and resistant to α-amylase and PAS-positive and breakable by α-amylase when more corsely grained. The latter therefore was referred to as glycogen by Davies et al. (1961a, b). It should be identical with the large glycogen areas we traced with the electron microscope (Fig. 20b). The function of this abundance of glycogen is not clear to us. But its connection with the forming of lacunae is conspicuous. It is also evident that prior to phases of lacunae formation, the abundance of glycogen is always especially pronounced.

The fine-grained α-amylase resistant material may be glycoproteids or mucopolysaccharides. Electron microscopical investigations correspondingly revealed a great amount of membrane bound electron-dense granules in the subplacental syncytium (Davies et al., 1961a, b). Their diameter varies, but hardly exceeds 0.5 μ (Fig. 20c). The content of the first of these granules was still very light when we discovered it on the 23rd day. Up to the 29th day there are only a few and their distribution is uneven. Only from the 32nd day onwards can vast amounts be found in the syncytium of the subplacenta. We could not find them in the other tissues of the subplacenta. From the 32nd day as well it can be shown that these granules are secreted into the lacunae. Here they seem to take part in what has been described above as deposition of electron-dense content of the lacunae (Fig. 20c). This is in accordance with the histochemical finding that the contents of the lacunae is also PAS-positive and resistant to α-amylase. It is conspicuous that these secretory granules subsist beyond the 55th day when significant regressive processes occur in the syncytium. Even on the 60th or 65th day when the remaining structures of the subplacenta have been destroyed, large amounts of them − now mostly without membrane − are still distinct in the decomposing subplacental region.

PAS-positivity and α-amylase resistance confirm the assumption that the fine-grained electron-dense structures within the granules are glycoproteins (Davies et al., 1961a, b; King and Tibbitts, 1976). Its secretion into the maternal lacunae indicates that it may be either a glycoprotein hormone or an enzyme. Attempts to define the identity of this secretion have been in vain so far. Also the purpose of this secretion is obscure since the granules are secreted into the maternal lacunae when the maternal circulation has already come to standstill. This means that the secretion cannot reach the maternal circulation.

Enzyme histochemistry has led to a number of interesting detail findings (Sax and Kaufmann, 1976), but to no decisive functional explanations (Table 13). The hexokinase proof is positive mainly in the subplacental syncytium up to the 58th day and becomes drastically weaker thereafter. Since the glucose-6-phosphatase proof in the

---

double membrane. After disintegration of the inner membrane and glycogen, the outer membrane limits a new-formed lacuna X 15300. (c) Detail micrograph from the 31st day. In the syncytoplasm there are several secretory granules. These are partly secreted (arrow) into the lacunae (*L*). It is not certain whether the high electron-density of the content of the lacunae can be explained by this secretory products, secreted by the syncytoplasm but not carried away by the maternal circulation. X 23400. Fixation (Fig. 17a−c): Immersion-fixation with 6 % glutaraldehyde, OsO_4-post-fixation

Table 13. Tabular comparison of the enzyme-histochemical findings during the development of the subplacenta. The data were obtained subjectively by comparing the interlobium to the subplacenta visually. m. p. = main placenta, s. p. = subplacenta.
4 – very strong, 3 – strong, 2 – medium, 1 – weak, (1) – doubtful positive, 0 – no detectable activity (according to findings by Sax and Kaufmann, 1976)

Enzyme	Age in days	20	30	40	50	56	58	63
Hexo-kinase	m.p.interlobium	4	4	4	4	4	4	4
	s.p. syncytium	4	4	4	3	3	0	0
	s.p. cytotroph.	2	2	2	3	3	0	0
	s.p. con. tissue	1	1	1	2	2	2	2
G6PDH		4	4	4	4	4	4	4
	,,	4	3	3	2	2	0	0
		1	3	3	3	3	0	0
		(1)	1	1	1	1	2	2
LDH		3	3	3	3	3	3	3
	,,	2	3	3	4	3	0	0
		1	1	2	2	3	0	0
		(1)	(1)	1	1	1	1	1
IDH		4	4	4	4	4	4	4
	,,	2	2	1	1	(1)	0	0
		1	1	1	1	(1)	0	0
		(1)	(1)	1	1	1	1	1
SDH		4	4	4	4	4	4	4
	,,	4	(1)	(1)	(1)	(1)	0	0
		2	2	2	2	2	0	0
		1	1	1	2	2	2	2
GDH		1	1	1	1	1	1	1
	,,	1	1	2	4	3	0	0
		3	3	3	3	3	0	0
		2	2	2	2	2	2	2
αGPDH		1	1	1	1	1	1	1
	,,	1	1	1	1	1	0	0
		3	3	3	3	3	0	0
		(1)	1	2	2	2	2	2
LAPase		(1)	(1)	(1)	(1)	(1)	(1)	(1)
	,,	(1)	(1)	(1)	(1)	(1)	1	2
		(1)	(1)	(1)	(1)	(1)	1	2
		4	4	4	4	4	4	4
acP		(1)	(1)	(1)	(1)	(1)	(1)	(1)
	,,	1	(1)	(1)	(1)	(1)	1	3
		1	(1)	(1)	(1)	(1)	(1)	1
		0	0	0	0	(1)	(1)	1

subplacental syncytium is almost negative at the time in question the conclusion may be drawn that much glucose is taken up here and phosphorylated but not passed on (cp. Stark and Kaufmann, 1973). Thus consumption of glucose seems to be very high in the subplacenta. This is in accordance with the high activity of glucose-6-phosphate dehydrogenase, for example, and lactate dehydrogenase in the subplacental syncytium. The high activity of lactate dehydrogenase, even exceeding that of the main placenta together with the very weak activity of the citrate-cycle enzymes, indicates that the major portion of energy is gained from glycolysis — possibly owing to deficient vascularization of the subplacenta. The high activity of glucose-6-phosphate dehydrogenase and the activity of glutamate dehydrogenase which is also very high in the second half of pregnancy are indicators of a strong ribose and protein metabolism; they may be related to the strong secretory performance of the subplacenta. The fact that the activity of aminopeptidase and acid phosphatase increases as from the 55th day whereas the activity of all other enzymes decreases markedly may be related to autolytic processes.

For the time being, statements on the function of the subplacenta of the guinea-pig can hardly be made. The special rôle of the subplacenta seems to be due to very early maternal vascularization which is not immediately followed by fetal vascularization in contrast to the main placenta. It may be possible that the involution of the maternal circulation is connected with the delayed beginning of the fetal circulation since this part of the developing placenta does not fulfil its purpose as a feto-maternal exchange organ. As the beginning regressive processes are confined to the maternal circulation and do not affect the subplacenta as a whole, it cannot be concluded that it is a more or less useless rudiment altogether (cp. Perrotta, 1959: Roberts and Perry, 1974). It cannot be said for sure whether the secretion which is the only positive process is of importance. Should the secretory products be enzymes, they may play a rôle in detachment of the placenta and the subsequent healing of the detachment area. Should they be glycoprotein hormones (Davies et al., 1961a, b; King and Tibbitts, 1976), they cannot be of any relevance for the progress of gestation because they do not reach the fetal circulation. This would also indicate its rudimentary character. It is also possible that the secretory products, having accumulated until the regression of the subplacenta thereafter reach the maternal and fetal circulation and assume a hormonal function in connection with the forthcoming birth.

6.5. Yolk Sac and Margin of the Placenta

Description of the guinea-pig placenta morphology would be incomplete without inclusion of the yolk sac. Many authors emphasize the great functional importance of the so-called yolk sac placenta for the metabolic processes between mother and fetus. (Asai, 1914; Anderson and Leissring, 1961; Leissring and Anderson, 1961; Larsen, 1963a, b; Padykula et al. 1966; Krzyzowska-Gruca and Schiebler, 1967; Carpenter and Ferm, 1969; King and Enders, 1970a, c). As obvious from a number of experimental researches, it plays an essential rôle in the earlier terms of pregnancy, when the main placenta is not yet adequately developed (see Dempsey, 1953; Anderson and Leissring, 1961). However, more recent researches have shown that the yolk sac has an essential

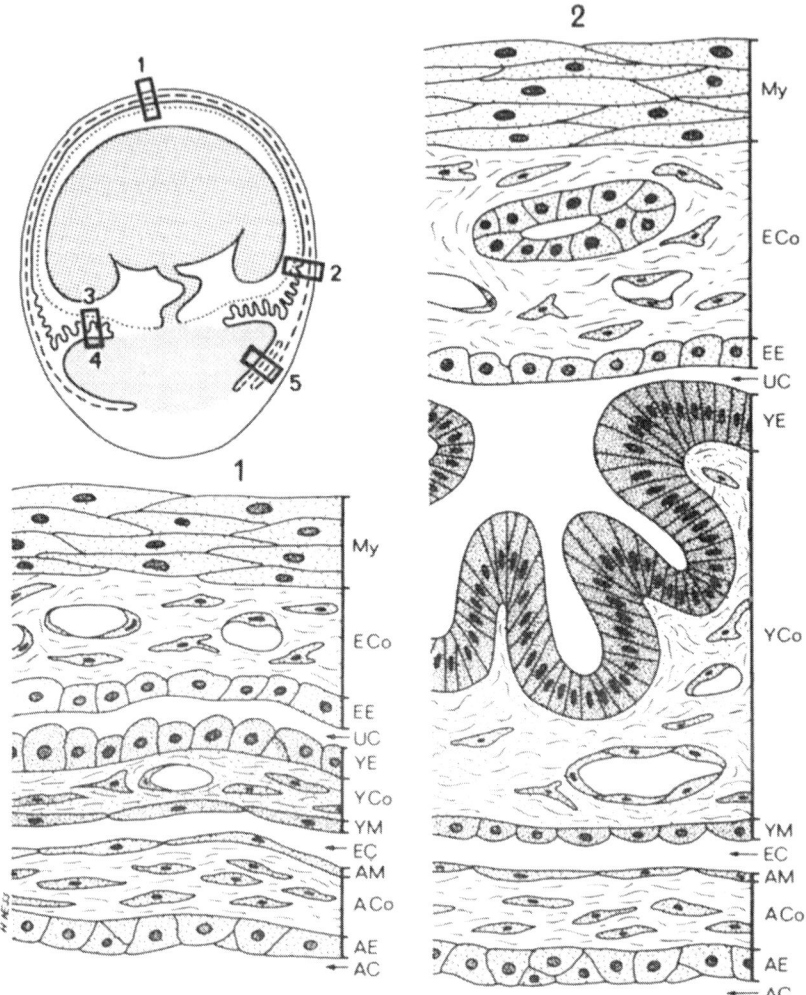

Fig. 21 (sections 1 and 2)

Fig. 21. Schematized representation of the different sections of the yolk sac. The origin of the section 1–5 is indicated in the survey sketch (top left). *Section 1*: Yolk sac far from the placenta, flanked by the uterus wall (above) at its epithelial surface, at its mesothelial surface (below) by the amnion. *Section 2*: The yolk sac near the placenta differs from section 1 by higher epithelium, more ample fetal vascularization and more pronounced formation of folds. *Section 3*: This section of the yolk sac detaches from the uterus wall to penetrate at the placental surface printing towards the fetus. At its mesothelial surface (above) it is covered by amnion. At its epithelial surface it is in contact with section 4 of the yolk sac. Structurally it largely corresponds to section 2. *Section 4*: As from section 4, the yolk sac cover of the placental surface pointing towards the fetus, the yolk sac epithelium lacks a basal layer of connective tissue with vessels. It is directly connected with the interlobium of the main placenta by a homogeneous membrane (membrane of Reichert) enclosing giant cells of Duval (remnants of cytotrophoblast). *Section 5*: The lateral yolk sac cover of the placenta corresponds in terms of structure to section 4, but has in contrast to the latter direct contact with the endometrium. In this region the endometrium forms large folds (remainders of the decidua capsularis?). The space between endometrial epithelium and yolk sac epithelium, as well as the space between the folds of the yolk sac epithelium have been depicted too

74

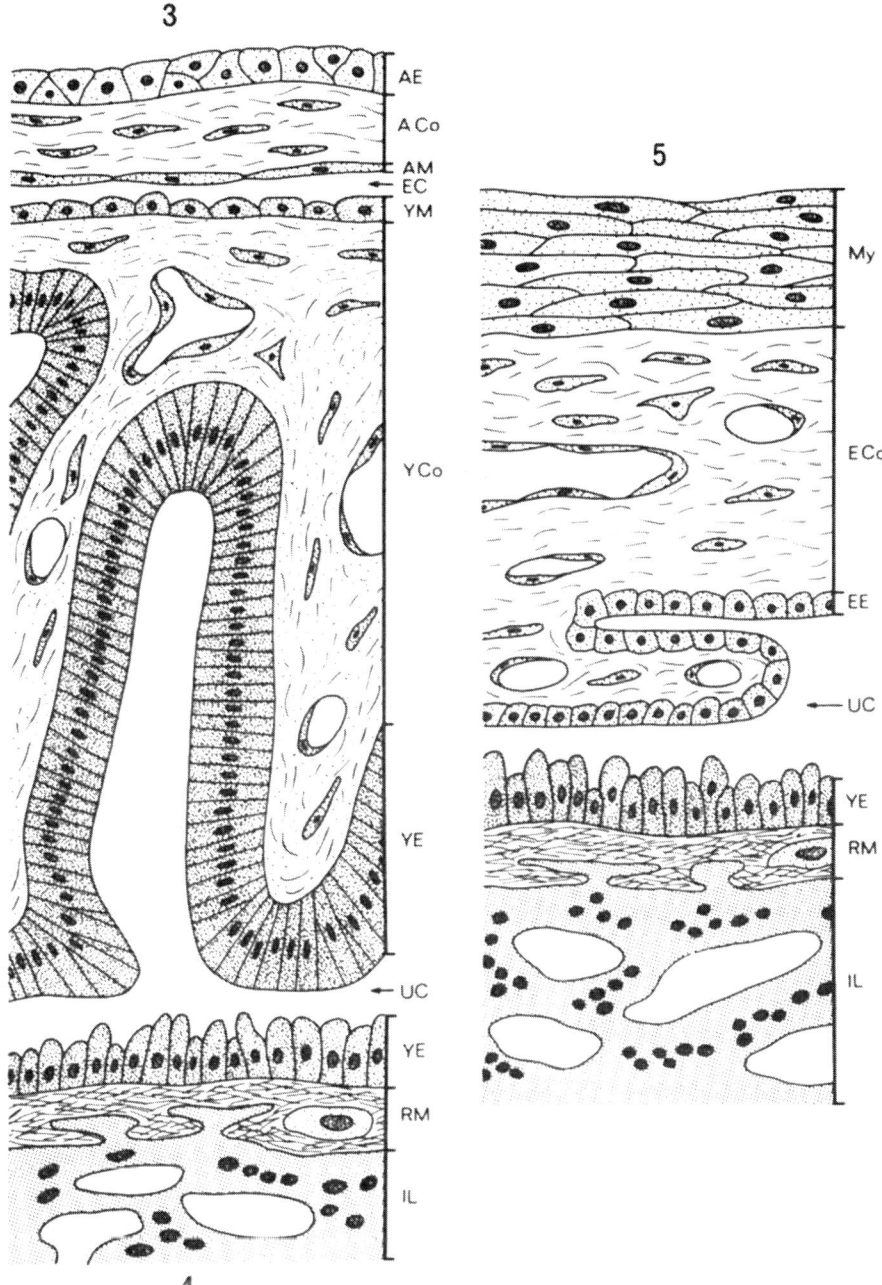

Fig. 21 (sections 3–5)

widely for the sake of clearness. The same applies to the space between amniotic and yolk sac mesothelium. All these spaces are in vivo capillary clefts.

Abbreviations used: My = myometrium; ECo = endometrial connective tissue; EE = endometrium epithelium; UC = uterine cavity; YE = yolk sac epithelium; YCo = yolk sac connective tissue; YM = yolk sac mesothelium; EC = exocoel; AM = amnion mesothelium; ACo = amnion connective tissue; AE = amnion epithelium; AC = amniotic cavity; RM = membrane of Reichert; IL = interlobium

practical bearing also on metabolic and transport processes developing during the later pregnancy periods (King and Enders, 1970c; King, 1972).

Following structure and vicinity relations, the yolk sac has to be subdivided into five sections. *Section 1* (Fig. 21a, b), the part furthest from the placenta (see Petry and Kühnel, 1963; Davidoff and Schiebler, 1970b) extends from the antiplacental pole of the fetal membranes up to 1 cm distance from the placenta. It is characterized by a largely smooth epithelial surface. At its outer epithelial surface it is flanked by the endometrium, at its mesometrial surface by the amnion. Both normally lie close to the yolk sac separated from it by a capillary cleft only.

Section 2 (Fig. 21a, c) is a stria of 1 cm width near to the placenta, also located between endometrium and amnion, whose epithelial surface has villi and foldings.

Section 3 (Fig. 21a, d) is, at term, a stria of approximately 1/2 cm width which detaches from the endometrium shortly before reaching the rim of the placenta, bends inwardly towards the umbilical cord an inserts at the placental surface pointing towards the fetus. Like section 2 ist has foldings and villi. At its mesothelial surface it may have contact with the amnion. More often than not fluid-filled widenings of the capillary cleft occur here which then are to be referred to as remnants of the exocel. Its epithelial surface no longer has contact with the endometrium but lines the epithelial surface of section 4.

Section 4 (Fig. 21a, d) is the part of the yolk sac which fuses with the marginal syncytium on the surface of the placenta pointing towards the fetus. It extends to the side-rim of the placenta and is mainly smooth.

Section 5 (Fig. 21a, e) is fused laterally and basally with the marginal syncytium of the surface of the placenta up to the rim of the placental stem. Its free epithelial and largely smooth surface is in contact with the endometrium.

The sections 1–3 whose epithelium convers a vascularized layer of connective tissue are referred to together as the visceral yolk sac placenta (Dempsey, 1953). This is distinguished from the parietal yolk sac placenta (King, 1972) (section 4 and 5) whose epithelium lines the placental surface together with the membrane of Reichert without connective tissue. The terms "visceral" and "parietal", however, we deem unfortunate since in the guinea pig the yolk sac epithelium as a whole stems from the visceral layer of the yolk sac epithelium (cp. chapter 3.2.). The parietal layer of the yolk sac is not formed in the guinea pig in contrast to the rat and mouse (cp. Grosser, 1927). The older term "ectoplacental endoderm" for sections 4 and 5 seems more appropriate.

Owing to differing vascularization and varying superficial contacts these five sections are certainly not functionally identical in spite of certain main similarities. Their morphology shall therefore be dealt with separately as far as necessary.

6. 5. 1. Section 1 of the Yolk Sac

Section 1, the part far from the placenta, consists of a onelayered stratum of mostly cubic cells; locally also there are flat or prismatic cells (Fig. 21b). They lie on a $5-10\ \mu$ thick layer of connective tissue which is limited by mesothelium and vascularized fetally. The epithelial cells are connected by desmosomes. The intercellular spaces are in particular basally widened to a marked degree. The basal plasmalemm is separated from the basal membrane in most places by labyrinthine foldings. The free surface of the epithelium is densely covered by short microvilli. These are usually in direct contact with the microvilli of the bordering parietal endometrial epithelium with which they interdigitate partly. The space between the two epithelia (the uterine cavity) is widened in places by extracellular electron-dense globulous masses with diameters up to 50 μ. It is not clear whether these are accumulated secretory products of the few endometrial glands or necrotical remainders of the decidua capsularis which can no longer be traced in late pregnancy.

The yolk sac epithelial cells contain a great many lysosomes which indicate intense transport of substances. Their organelle pattern otherwise corresponds to that of the yolk sac epithelial cells near the placenta (section 2, see below). A very thin basal membrane establishes the connection to the stroma of connective tissue consisting of nets of collagen fibres with 1–2 layers of fibroblasts. Small fetal vessels are widely scattered even at the anti-placental pole. They are connected with the umbilical cord via the stem of the yolk sac (remnants of the connective stalk). Towards the amnion the stroma is limited by an extremely thin layer of mesothelium. This is in contact with the amniotic mesothelium through a cleft which in-vivo only has capillary width. The two mesothelia may disappear locally when the yolk sac mesoderm and the amnion mesoderm fuse. An open exocel filled with fluid is not existent here.

The amnion is not vascularized. The thin amniotic mesothelium is followed by a layer of connective tissue 10–20 μ thick, consisting of collagen fibres and flattened fibroblasts. It is covered by the strongly thinned approximately 5 μ thick amniotic epithelium on the side towards the fetus. There are no conspicuous signs of metabolic activity ultrastructurally or histochemically.

Should transport processes play a major rôle here, then certainly it would be only from the richly vascularized endometrium through the endometrial and yolk sac epithelium into the yolk sac vessels or vice versa. The markedly widened intracellular spaces of both epithelia may be an indication for intensive fluid exchange.

6. 5. 2. Section 2 of the Yolk Sac

The part of the free yolk sac close to the placenta (Fig. 21c) differs already light-microscopically from the part far from the placenta. The epithelial surface is strongly enlarged partly by slim folds, partly by villous growths of 200–300 μ in length. Only the flattened tips of the villi and folds come into contact with the endometrial epithelium. The space between the two epithelia (uterine cavity) is filled by homogenous masses of varying stainability of an evidently clotted exsudate. The layer of connective tissue of the yolk sac is 40–50 μ thick, shows abundance of cells and is richly vascularized. It is limited by a cubic mesothelial layer towards the amnion. Except for a short length at the transition to section 3 it is everywhere in direct contact with the amnion. The amnion itself is in this region definitely thicker than in the part far from the placenta due to a thicker layer of connective tissue, but otherwise is structurally unchanged.

Depending on gestational age, electron-microscopy and histochemistry render differing functional images of the yolk sac epithelium. Yet on the 14th day (Davidoff and Schiebler, 1970b), the structure of the part of the yolk sac near the placenta is characterized by the presence of high columnar cells, disposed over a comparatively broad basement membrane which broadens with the progress of pregnancy (King and Enders, 1970a). The cell membranes in their apical part are connected with each other through desmosomes, whereas on the lateral aspects they form intercellular spaces varying in width wherein cytoplasmic protrusions reside. In the apical portion of the cells there are numerous microvilli, part of them with distended endings. With the advance of pregnancy they become equally wide along their full course (see Dempsey, 1953).). Between the microvilli in the direction of cytoplasm, invaginations of varying depth are formed, dividing in the form of pinocytotic bubbles and vacuoles in the cytoplasm. The nuclei are located mainly at the base of the cells, and components of the granular endoplasmic reticulum are situated around them. We came across many ribosomes and polyribosomes. The mitochondria in earlier stages are with a brighter matrix but, from about the 20th day onwards, they become electron dense. In this period, the numerous lipid drops of varying size and density, and mainly basal localization, showing a marked decrease with the progress of pregnancy are most impressive (see King and Enders, 1970a). In the later stages, the granular endoplasic reticulum augments (Dempsey, 1953), whilst the quantitiy of free ribosomes decreases. In the early terms the connective tissue stroma bears a large numer of different mesenchymal cells in the cytoplasm of which around the 20th day once again numerous lipid drops are established. Subsequently, their numer diminishes whereas the connective tissue fibres show an increase. Comparatively early, in the yolk sac stroma adequately developed vitelline vessels are observed.

In the early gestation terms, the yolk sac is one of the components exhibiting a strongly pronounced histochemical activity for a number of enzymes. Both the ecto-placental endoderm, and the part close to the placenta show an intense HSTD activity (Vollrath, 1965; Davidoff and Gospodinov, 1971) which declines about the 37th–45th day. This is an indication that in this particular period, the yolk sac constituents most probably participate in the steroid metabolism. The demonstration of strong LAPase and GDH (Vollrath, 1965) points to the rôle of epithelial cells in the protein transport and metabolism. This is furthermore supported by the data of Barnes (1959), Anderson and Leissring (1961) and Leissring and Anderson (1961) concerning the ability of the yolk sac to absorb and transport serum proteins (antibodies) from mother to fetus, mainly in the first half of pregnancy. The presence of glycogen, G6Pase and PAS-positive staining in the apical part of the cells emphasizes their rôle in carbohydrate metabolism. The high SDH (Reale and Pipino, 1959; Vollrath, 1965), CO, LDH, G6PDH, $NADH_2$-Red and $NADPH_2$-Red activity is interpreted as evidence of active energetic processes, whereas βHBDH activity démonstration may be related to lipid metabolism (Vollrath, 1965).

During the second half of gestation, the yolk sac epithelium is presented electron-microscopically by columnar cells (Dempsey, 1953; Petry and Kühnel, 1963; Davidoff and Schiebler, 1970b; King and Enders, 1970a). The apical zones of the cells possess many microvilli which may also contain filamentous material. The membranes of adjacent cells are connected through desmosomes in the apical part. At some points, the intercellular spaces are dilated. Often, within the distension zone, along the apical membrane and along the membrane at the base of the cells, coated vesicles are detached. Abundance of rounded structures occur in the apical cytoplasm and in the supranuclear zone (endocyte vacuoles, dense droplets, lysosomelike bodies), as well as Golgi zones and of variable size mitochondria. Granular endoplasmic reticulum is established perinuclearly and basally. Glycogen granules are seen throughout the cytoplasm. The nuclei show basal localization with mainly peripherally positioned chromatin.

Histochemically, in the apical part of the above cells a marked alkP activity (Hard, 1946; Petry and Kühnel, 1963), as well as large glycogen quantities (Wislocki et al.,

1946; Davies, 1956; Petry and Kühnel, 1963) are established. The presence of numerous microvilli, endocyte vesicles and other vacuoles prove the powerful capacity of epithelial cells to absorb substances coming from the uterine cavity. Researches by King and Enders (1970a) have shown that the epithelial cells may absorb peroxidase and ferritin and transport them to the vitelline vessels or mesothelium, unlike thorotrast which is absorbed to a lesser degree, and not transported. This points to a selectivity in the absorptive function of the yolk sac columnar epithelium. Serum proteins (antibodies) in the second half of pregnancy also may be absorbed by the epithelial cells and transported to the fetus (Anderson and Leissring, 1961; Leissring and Anderson, 1961) which proves the yolk sac participation in the immunological processes. A great part of oxidative and hydrolytic enzymes reduce their activity considerably during the second half of pregnancy which suggests that the yolk sac has a smaller part in the total exchange between mother and fetus.

The yolk sac basement membrane is $1.5-2.0 \mu$ wide (Dempsey, 1953; Petry and Kühnel, 1963; Davidoff and Schiebler, 1970b; Davidoff, 1970; King and Enders, 1970a) and is made up of homogenous to finely granulated or filamentous material. In the membrane numerous rounded bright spaces are observed, imparting to the former a lacy appearance. Petry and Kühnel (1965) explain this basement membrane structure by peculiarities of the protein transport through it in the rabbit.

The yolk sac stroma is composed of different connective-tissue cells and fibers. Between the cells fibrocytes, fibroblasts, macrophages and a great number of undifferentiated mesenchymal cells are distinguishable with polymorphism being characteristic for all of them (King and Enders, 1970a; Davidoff, 1970). Vitelline vessels are localized between the elements of the stroma. According to King and Enders (1970a), after the 26th gestation day certain segments of their endothelium are fenestrated. A thin mesothelial layer separates the connective-tissue stroma from the exocoel following an increase of the collagen fibers towards the mesothelium (see King and Enders, 1970a).

6. 5. 3. Section 3 of the Yolk Sac

Section 3 (Fig. 21d) of the yolk sac differs structurally only insignificantly from section 2. Villi and folds are mostly flatter, broader and partly ramified. There are no differences in vascular supply and ultrastructure. However, between the villi as well as in the space under the endoderm cover of the placental surface (section 4) there are no clotted masses of exsudate to be found as on the surface of section 2. Since ultrastructure and histochemistry do not give any hint of lesser metabolic activity than in the preceeding part, the question as to the function of this section arises. The villi of the yolk sac cling so closely to the ectoplacental endoderm on the surface of the placenta that an important fluid flow from the endometrium to this region or vice versa seems improbable. Transport from the main placenta to the yolk sac or vice versa would be pointless as both are connected to the same vascular system. Since special experimental investigations on this region still remain to be done, its rôle is still unclear for the time being.

6. 5. 4. Section 4 of the Yolk Sac

Section 4 (Fig. 21d) arises from section 3 following the turn of the yolk sac epithelium from the free yolk sac on to the placental surface. Thereby the epithelium loses its stroma of connective tissue and the fetal vessels therein. It is now directly attached to the membrane of Reichert which arises from the fusion of the basal membrane of the yolk sac with that of the marginal trophoblast. In the very first stages of development it is lined inwardly (towards the placenta) by a complete layer of trophoblastic cells which establishes the connection to the marginal syncytium. When this cytotrophoblastic layer recedes between the 20th and 30th day, its remaining cell elements grow bigger (giant cells of Duval). Concomitantly they get sheated by the membrane of Reichert which increases in thickness. In the second half of gestation the giant cells of Duval are only to be found now and then – usually at the rim of the placenta and near the placental stem (Kaufmann, 1969a). The membrane of Reichert now reaches a thickness of 10 μ up to 50 μ exceptionally. It is often interrupted by processes of the giant cells and of the underlying marginal syncytium. Light-microscopically the yolk sac epithelium of this region is much more basophilic than the epithelium of the free yolk sac. Ultrastructurally there are hardly any differences except for a lesser number of lysosomes.

6. 5. 5. Section 5 of the Yolk Sac

Section 5 (Fig. 21e) is mainly characterized by the fact that the ectoplacental endoderm again comes into contact with the endometrium at the transition to the lateral parts of the placenta. Large, flat endometrial folds usually not mentioned in publications until now are inserted here between uterus wall and placenta; they sometimes lie in two or three layers. As these folds as well are richly vascularized maternally they do not present hindrance to transport. They may possibly be remainders of the decidua capsularis.

Ultrastructurally the endodermal cells of this and the previous region do not differ significantly from those in the sections of the free yolk sac: The cells are situated over the broad Reichert's membrane (Wislocki and Padykula, 1953) whose side facing them shows uneven contours, but the one facing the trophoblast is smooth. The membrane is made of electron-bright, homogeneous to slightly fine-granular mass. Beneath the Reichert's membrane fragments of varying length from the marginal trophoblast are established (King, 1972) whose ultrastructure was already described in the main placenta interlobe. Apart from the trophoblast, under Reichert's membrane here and there giant cells of Duval are detected which, after the 32nd gestation day undergo a reverse development (Kaufmann, 1969a) and contain an abundance of degenerative structures in their cytoplasm. By the 35th day, Davidoff and Gospodinov (1971) observed a strongly pronounced HSTD activity in the ectoplacental endoderm and in the giant cells, considered as a likely outward sign of the participation of these cells in the steroid metabolism. Contrary to the visceral part of the yolk sac, here King (1971, 1972) found a weaker absorption and an absence of transport of exogenous peroxidase and ferritin from the endodermal cells. Probably the particles absorbed at this level undergo decomposition in the lysosome system. However, in the parietal yolk sac, the transport of peroxidase and ferritin in an opposite direction, e. g. from the maternal lacunae through the trophoblast, Reichert's membrane and intercellular spaces towards the lumen of the uterus, is effected very rapidly. The pathway outlined is discussed by the author as a channel of outmost importance for the accomplishment of the mother-to-fetus transport of substances. The rôle played by the components of the parietal yolk sac placenta is still not fully studied.

7. Possibilities of Comparison with the Human Placenta

A strong incentive for morphological research on the guinea-pig placenta was the hope of finding an easily available model for the human placenta since experimental studies are hardly possible on the latter. Consequently the question arises as to what extent the placenta of the guinea-pig can be compared to that of man.

A decisive advantage of the cavia-placenta compared to that of all other test animals lies in the relatively long gestation of guinea-pigs, lasting 63 days or more. The morphological details of the guinea-pig placenta also bear most similarities to that of the human; yet discrepancies are so great that transferable statements may not be quite unproblematical.

The main placenta of the guinea-pig is largely hemomonochorial like that of man. There are, however, two very significant differences due to the fact that the human placenta is villous and that of the guinea-pig lacunal:

1) In lacunal placentae the blood stream has a clearly defined direction and follows the counter-current principle. In the villous placentae on the other hand, there are blood-stream directions which are hardly definable and which are termed the pool-flow-system. Since all transport processes in the placenta are not only influenced by the nature of the membranes of the trophoblast but also by the blood-stream conditions, and since both parameters are difficult to consider separately, the physiological transport data obtained on pregnant guinea-pigs can only be applied restrictedly to man.

2) The villous structure necessitates the existence of an extensive stroma of connective tissue in the villi. This forms the preponderant volume component of the organ in the case of man. The proportion of connective tissue in the guinea-pig placenta, however, is minimal. According to histochemical and experimental investigations the function of the connective tissue is not only mechanical, but influences feto-maternal metabolic relations to an extent not yet known. As to the amount of connective tissue, there are such important differences that any comparative data founded on weight units of the placenta have no firm base. The mean placental weight of the guinea-pig at term, for instance, amounts to 5 g to 85 g mean embryonic weight. This placento-fetal weight relation of 1:17 is opposed to one of 1:6 in man. Only superficial considerations would allow the conclusion that the guinea-pig placenta is three times as efficient as that of man due to its ideal circulatory conditions. The dubiousness of this assumption becomes apparent when regarding the two placentae without the connective tissue, which is not irrelevant in terms of metabolism. One then obtains practically the same weight relations in man and guinea-pig.

In spite of considerable differences in the rough morphology, there are numerous comparable details in the fine structure. The ultrastructural research of the main placenta reveals zones of different structure and histochemistry: interlobium, transitional zone, periphery of the lobe, middle of the lobe, centre of the lobe. These regions are characterized by typical enzyme patterns and, as shown above, are open to rough functional interpretation. The villous surface of the human placenta as well is organized in areas differing structurally. These regions disposed in a mosaic-like way, are so small and are delimited so imprecisely that their structural and even more their histochemical evaluation presents difficulties. We then focussed our hopes on the structural and histochemical comparison of the two placentae which led to the discovery and the interpretation of a number of comparable structures (Kaufmann, 1974; Kaufmann et al., 1974). The epithelial plates of the human placental villi correspond largely to the syncytoendothelial membranes forming the main part of the periphery of the lobe in the guinea-pig. These areas chiefly serve the exchange of small-molecular substances and the transport of glucose. The thicker sections of the surface of the villi in the human placenta, devoid of nuclei or poor in nuclei, correspond in the main to the centre of the lobe. Mainly active transport, together with processes of decomposition and transformation (protein metabolism) take place here. The syncytial knots show similarity with the transitional zone from the interlobium to the labyrinth. The proliferation knots resemble in many aspects the interlobium structurally and histochemically. As great as the similarities of the latter structures may be, comparability is questionable in their case. Syncytial and proliferation knots in the human placenta no longer count uncontestably today as sites of proliferation but as places of ageing trophoblast caused by shifts of nuclei. Shortly before its definite destruction however it attains a maximum of activity in energy- and nucleic-acid-metabolism. The interlobium and the transitional zone of the guinea-pig placenta are centres of such energy- and nucleic-acid-metabolism, but contrarily to their "human counterparts" they are by no means sites of degeneration, but rather are centres of proliferation and increase for the labyrinth. Here also we find a superficially functional similarity in largely differing formations, which shows how risky it is to deduce overall comparability from a few similarities.

Regarding subplacenta and fetal membranes, there are even fewer similarities. There is no counter-part for the subplacenta in the human placenta. Also the correspondingly located basal plate does not bear the slightest resemblance to the subplacenta in terms of structure, histochemistry or function. The outer fetal membrane in the guinea-pig — the yolk sac — is in a position corresponding to that of the chorion laeve in man. There are also certain structural conformities. The fundamentally different origin of the two — the yolk sac as an endodermal formation of the fetus, the chorion laeve as derivative of the trophoblast — destroy any hope of common identity. The chorion laeve has to be considered more a restrictedly functioning receding product of the placenta, the yolk sac a highly active metabolic membrane throughout gestation.

This estimation which seems negative in many aspects is in the main of importance for the placenta morphologist only. For all physiological, pharmacological and pathological investigations, placentae of primates may be more appropriate. However, since they are hardly ever at disposal, the placenta of the guinea-pig has, in spite of all reservations mentioned, most in common with the human placenta compared with that of any other available test animal. Both organs — the human and the guinea-pig placenta — have matured in different ways from only slightly comparable formative tissues to formations which are only partly similar in terms of structure, but ultimately equivalent functionally. Both are hemomonochorial, meaning that the placental barrier, which is of so great importance for all pharmacological and physiological transport studies, has a similar structure (cp. Wilde et al., 1946; Fuchs and Fuchs, 1957a, b; Dancis, 1960; Folkart et al., 1960; Levitz, 1960; Twardock, 1967; Douglas et al., 1968; Reynolds and Young, 1971; Schröder et al., 1972; Carstensen et al., 1973; Hill and Young, 1973; Wong and Morgan, 1973; Schröder et al., 1975). In both a syncytially fused trophoblast is in direct contact with the blood. This allows the assumption — but not proof — that the syncytotrophoblast of both also tends to similar biological reactions and ways of functioning. To the extent to which experimentally obtained results can be attributed to functions of this trophoblast, these results may be transferable to the human placenta.

Acknowledgements

We would like to thank Mrs. E. Böhm, Mrs. E. Schäfer and Mr. R. Franck for excellent technical help, Miss H. Klein and Mr. H. Hess for skilful photographic and artistic work as well as Mr. H. Ardill, Mr. W. Kaiser and Mr. K. Meleksitian for the english translation and Mrs. S. Pancic for untiring help in writing the manuscript.

The investigations for this publication were carried out in part by the authors individually at the Anatomical Institute, University of Hamburg and at the Regeneration Research Laboratory, Bulgarian Academy of Sciences, Sofia, respectively. Another part of the work was done together at the Anatomical Institute, University of Würzburg. We are indebted to Prof. Dr. T. H. Schiebler for the use of the material won at this institute.

8. References

Amoroso, E. C.: Placentation. In: Marshall's, Physiology of reproduction. London: Longmans, Green & Co. 1952

Anderson, J. W., Leissring, J. C.: The transfer of serum proteins from mother to young in the guinea pig. II. Histochemistry of tissues involved in prenatal transfer. Amer. J. Anat. 109, 157–174 (1961)

Asai, T.: Zur Entwicklung und Histophysiologie des Dottersacks der Nager mit Entypy des Keimfelds. Anat. Hefte 51, 467–641 (1914)

Bailey, D. J.: Counter-current flow of maternal and fetal bloodstreams of guinea pig placenta. J. Physiol. (Paris) 242, 104P (1974)

Barnes, J. M.: Antitoxin transfer from mother to foetus in the guinea pig. J. Path. Bact. 77, 371–380 (1959)

Bartels, H., Yassin El, D., Reinhardt, W.: Comparative studies of placental gas exchange in guinea pigs, rabbits and goats. Resp. Physiol. 2, 149–162 (1967)

Bischoff, Th. L. W.: Entwicklung des Meerschweinchens. Diss. Univ. Giessen 1852

Bischoff, Th. L. W.: Neue Beobachtungen zur Entwicklungsgeschichte des Meerschweinchens. Abh. Bayr. Akad. Wiss. Math. Phys. Kl. X, München 1866

Bjellin, L., Sjöquist, P.–O. B., Carter, A. M.: Uterine, maternal placental and ovarian blood flow throughout pregnancy in the guinea pig. Z. Geburtsh. Perinat. 179, 179–187 (1975)

Blandau, R. J.: Observations on implanation of the guinea pig ovum. Anat. Rec. 103, 19–47 (1949a)

Blandau, R. J.: Embryo-endometrial interrelationship in the rat and guinea pig. Anat. Rec. 104, 331–359 (1949b)

Brunings, E. A., de Priester, W.: Effect of mode of fixation on formation of extrusions in the midgut epithelium of Calliphora. Cytobiology 4, 487–491 (1971)

Carpenter, S. J., Ferm, V. H.: Uptake and storage of thorotrast by the rodent yolk sac placenta: An electron microscopic study. Amer. J. Anat. 125, 429–456 (1969)

Carstensen, M., Leichtweiß, H. P., Schröder, H.: Zellpotentiale und Natrium-Kaliumverteilung in der Meerschweinchenplacenta. Arch. Gynäk. 215, 305–313 (1973)

Caulfield, J. B.: Effects of varying the vehicle for OsO_4 in tissue fixation. J. biophys. biochem. Cytol. 3, 827–829 (1957)

Christie, G. A.: Comparative histochemical distribution of glycogen and alkaline phosphatase in the placenta. Histochemie 9, 93–116 (1967)

Christie, G. A.: Comparative histochemical studies on carbohydrate, lipid and RNA metabolism in the placenta and fetal membranes. J. Anat. (Lond.) 103, 91–112 (1968)

Creighton, C.: On the formation of the placenta in the guinea pig. J. Anat. Physiol. 12, 534–590 (1878)

Dallam, R. D.: J. Histochem. Cytochem. 5, 178 (1957) cit.: Pearse, A. G. E.: Histochemistry, Theoretical and Applied Vol. I. London: Churchill Ltd. 1968

Dancis, J.: In: The placenta and fetal membranes (ed. C. A. Villee). New York: Williams & Wilkins Co., 1960

Dancis J.: The perfusion of guinea pig placenta in situ. Fed. Proc. 23, 781–784 (1964)

Dancis, J., Brenner, M. A., Money, W. L.: Some factors affecting the permeability of guinia pig placenta. Amer. J. Obstet. Gynec. 84, 570–576 (1962)

Dancis, J., Money, L.: Transfer of sodium and iodo-antipyrine across guinea pig placenta with an in situ perfusion technique. Amer. J. Obstet. Gynec. 80, 215–220 (1960)

Davidoff, M.: Elektronenmikroskopische Charakteristik und Histochemie der reifen Meerschweinchenplacenta und der Placenta während der Entwicklung. Diss. (bulg.) Sofia 1970

Davidoff, M.: The guinea pig placenta: fine structure and development. Acta anat. (Basel) 86, supp. 23–46 (1973)

Davidoff, M. Gospodinov, Chr.: Histochemische Untersuchungen über die Aktivität einiger Hydroxisteroiddehydrogenasen in der Meerschweinchenplacenta während der Entwicklung. Histochemie 28, 198–204 (1971)

Davidoff, M., Schiebler, T. H.: Über den Feinbau der reifen Meerschweinchenplacenta. Z. Anat. Entwickl.-Gesch. **130**, 216–233 (1970a)

Davidoff, M., Schiebler, T. H.: Über den Feinbau der Meerschweinchenplacenta während der Entwicklung. Z. Anat. Entwickl.-Gesch. **130**, 234–254 (1970b)

Davies, J.: Histochemistry of the rabbit placenta. J. Anat. (Lond.) **90**, 135–142 (1956)

Davies, J., Dempsey, E. W., Amoroso, E. C.: The subplacenta of the guinea pig, an electron microscope study. J. Anat. (Lond.) **95**, 311–324 (1961a)

Davies, J., Dempsey, E. W., Amoroso, E. C.: The subplacenta of the guinea pig: development, histology and histochemistry. J. Anat. (Lond.) **95**, 457–473 (1961b)

De Duve, C., Wattiaux, R.: Functions of lysosomes. Ann. Rev. Physiol. **28**, 435–492 (1966)

Dempsey, E. W.: Electron microscopy of the visceral yolk sac epithelium of the guinea pig. Amer. J. Anat. **93**, 331–364 (1953)

Diczfalusy, E., Mancuso. S.: Oestrogen metabolism in pregnancy. In: Foetus and placenta (eds. A. Kloper and E. Diczfalusy), p. 191–248. Oxford and Edinburgh: Blackwell Sc. Publ. 1969

Disse, J.: Die Eikammer bei Nagern, Insektivoren und Primaten. Ergebn. Anat. Entwickl.-Gesch. **15**, 530–580 (1905)

Douglas, T. A., Renton, J. P., Wright, R.: Role of transferrin in the placental transfer of iron in the rabbit. Amer. J. Obstet. Gynec. **102**, 1169–1172 (1968)

Draper, R. L.: The prenatal growth of the guinea pig. Anat. Rec. **18**, 369–392 (1920)

Duval, M.: Le placenta des rongeurs, II. De l'inversion des feuillets chez les rongeurs. J. Anat. (Paris) **26**, 521–592 (1890)

Duval, M.: Le placenta des rongeurs. III. Le placenta de la souris et du rat. J. Anat. (Paris) **27**, 24–96; 344–395; 515–612 (1891)

Duval, M.: Le placenta des rongeurs. IV. Le placenta du cochon d'Inde. J. Anat. (Paris) **28**, 58–408 (1892)

Egund, N., Carter, A. M.: Uterine and placental circulation in the guinea pig: An angiographic study. J. Reprod. Fertil. **40**, 401–410 (1974)

Enders, A. C.: A comparative study of the fine structure of the trophoblast in several hemochorial placentas. Amer. J. Anat. **116**, 29–68 (1965)

Enders, A. C., King, B.: The cytology of Hofbauer cells. Anat. Rec. **167**, 231–252 (1970)

Enders, A. C., Schlafke, S.: Cytological aspects of trophoblast-uterine interaction in early implantation. Amer. J. Anat. **125**, 1–30 (1969)

Ercolani, G. B.: Sull' unita del tipo anatomico della placenta nei mammiferi. Bologna 1879 Cit.: Duval (1892)

Faber, J., Hart, F. M.: The rabbit placenta as an organ of diffusional exchange. Comparison with other species by dimensional analysis. Circulat. Res. **19**, 816–833 (1966)

Fawcett, D. W.: Atlas zur Elektronenmikroskopie der Zelle. München-Berlin-Wien: Urban und Schwarzenberg, 1973

Ferguson, M. M., Christie, G. A.: Distribution of hydroxysteroid dehydrogenases in the placentae and foetal membranes of various mammals. J. Endocr. **38**, 291–306 (1967)

Fischer, W. M.: Das Strombahnsystem und der Austausch der Atemgase in der Meerschweinchenplazenta. Verh. d. Anat. Ges. **62**. Vers. Anat. Anz. 121 Suppl. 241–248 (1968)

Fischer, W. M., Vogel, H. R., Thews, G.: Der Saure-Basenstatus und die CO_2-Transportfunktion des mutterlichen und fetalen Blutes zum Zeitpunkt der Geburt. Pflügers Arch. ges. Physiol. **286**, 220 (1965)

Folkart, G. R., Dancis, J., Money, W. L.: Transfer of carbohydrates across guinea pig placenta. Amer. J. Obstet. Gynec. **80**, 221–223 (1960)

Franke, H. F.: Feinstruktur der Placenta. Jena: VEB Gustav Fischer 1969

Fuchs, F.: The red cell volume of the maternal and foetal vessels of the guinea pig placenta. Acta physiol. scand. **28**, 162–171 (1953)

Fuchs, F., Fuchs, R.: Studies on the placental transfer of phosphate in the guinea pig. Acta physiol. scand. **38**, 391–397 (1957a)

Fuchs, F., Fuchs R.: Studies on the placenta and their uptake of radioactive phosphorus. Acta physiol. scand. **39**, 277–285 (1957b)

Gerebtzoff, M.-A.: Nouvelles recherches histochimiques sur l'acetylcholinesterase dans le placenta de cobaye. Ann. Histochim. **2**, 3–10 (1957)

Goutier-Pirotte, M., Gerebtzoff, M.–A.: L'acetylcholinesterase dans le placenta du cobaye. Premiers resultats de recherches histochimiques. Arch. int. Physiol. Biochem. 63, 445–457 (1955)

Grosser, O.: Vergleichende Anatomie und Entwicklungsgeschichte der Eihäute und der Placenta. Wien und Leipzig: Wilhelm Braumüller 1909

Grosser, O.: Frühentwicklung, Eihautbildung und Placentation des Menschen und der Säugetiere. In: Deutsche Frauenheilkunde. Geburtshilfe, Gynäkologie und Nachbargebiete in Einzeldarstellungen. München: J. F. Bergmann 1927

Hard, W. L.: A histochemical and quantitative study of phosphatase in the placenta and fetal membranes of the guinea pig. Amer. J. Anat. 78, 47–78 (1946)

Hensen, V.: Beobachtungen über die Befruchtung mit Entwicklung des Kaninchens und Meerschweinchens. Z. Anat. Entwickl.-Gesch. 1, 212 (1876)

Hensen, V.: Ein frühes Stadium des im Uterus des Meerschweinchens festgewachsenen Eies. Arch. Anat. Physiol., Anat. Abt. 60–70 (1883)

Herrmann, E., Stolper, L.: Ein Beitrag zur Entwicklung des Meerschweinchens. Verh. dtsch. Gynek. Gesch. 1903 (1904)

Herrmann, E., Stolper, L.: Zur Syncytiogenese beim Meerschweinchen. Sitzungsber. d. Kais. Akad. d. Wiss. Math. naturwiss. Kl. Abtl. III 114 (1905)

Hill, P. M. M., Young, M.: Net placental transfer of free amino acids against varying concentrations, J. Physiol. 235, 409–422 (1973)

Hopsu, V. K., Ruponen, S., Talanti, S.: Leucine aminopeptidase in the placenta of the rat. Acta histochem. (Jena) 12, 305–309 (1961)

Ibsen, H. L.: Prenatal growth in guinea pigs with special reference to environmental factors affecting weight at birth. J. exp. Zool. 51, 51–91 (1928)

Janiak, I.: Das endokrine System bei der Fortpflanzung der Versuchs- und Nutztiere und des Menschen. Hannover: Verlag M. & H. Schaper 1971

Jollie, W.: Fine structural changes in placental labyrinth of the rat with increasing gestational age. J. Ultrastruct. Res. 10, 27–47 (1964)

Kaiser, W., Kaufmann, P.: Die Ultrastruktur der reifen Meerschweinchenplacenta in Abhängigkeit von Fixationsmittel und Fixationsmethode (in prep. 1977)

Kaufmann, P.: Die Meerschweinchenplacenta und ihre Entwicklung. Z. Anat. Entwickl.-Gesch. 129, 83–101 (1969a)

Kaufmann, P.: Über polypenartige Vorwölbungen an Zell- und Syncytiumoberflachen in reifen menschlichen Placenten. Z. Zellforsch. 102, 266–272 (1969b)

Kaufmann, P.: Untersuchungen über die Langhanszellen in der menschlichen Placenta. Z. Zellforsch. 128, 283–302 (1972)

Kaufmann, P.: Die Morphologie der Meerschweinchenplacenta. Vortrag auf dem Perinatologenkongress, Berlin 1973

Kaufmann, P.: Die Morphologie der Meerschweinchenplacenta nach Monojodacetat- und Fluoridvergiftung: Experimente zur Infarktgenese. Vortrag auf dem Perinatologenkongress, Berlin 1973b

Kaufmann, P.: Perfusionshistochemische und experimentelle Untersuchungen zum Energiestoffwechsel der Placenta. Vortrag auf der 68. Versammlung der Anatomischen Gesellschaft, Lausanne 1973. Verh. anat. Gesch. (Jena) 68, 283–287 (1974)

Kaufmann, P.: Über die Bedeutung von Plasmaprotrusionen an reifenden und alternden Zellen. Verh. anat. Ges. (Jena) 69, 307–312 (1975a)

Kaufmann, P.: Experiments on infarct genesis caused by blockage of carbohydrate metabolism in guinea pig placentae. Virchows Arch. path. Anat. 368, 11–21 (1975b)

Kaufmann, P., Schiebler, T. H., Ciobotaru, C., Stark, J.: Enzymhistochemische Untersuchungen an reifen menschlichen Placentazotten. II. Zur Gliederung des Syncytiotrophoblasten. Histochemistry 40, 191–207 (1974)

Kaufmann, P., Thorn, W., Jenke, B.: Die Morphologie der Meerschweinchenplacenta nach Monojodacetat- und Fluorid-Vergiftung. Arch. Gynäk. 216, 185–203 (1974)

Kayden, H.: Transfer of lipids across the guinea pig placenta. Exerpta. med. (Amst.), Series 170, 13 (1968)

Kayden, H., Dancis, J., Money, W. L.: Transfer of lipids across the guinea pig placenta. Amer. J. Obstet, Gynec. 104, 564–572 (1969)

King, B. F.: Differentiation of parietal endoderm cells of the guinea pig yolk sac, with particular reference to the development of the endoplasmic reticulum. Develop. Biol. **26**, 547–559 (1971)

King, B. F.: The permeability of the guinea pig parietal yolk sac placenta to peroxidase and ferritin. Amer. J. Anat. **134**, 365–376, (1972)

King, B. F., Enders, A. C.: Transport of horseradish peroxidase by the guinea pig visceral yolk sac. Anat. Rec. **166**, 331 (1970a)

King, B. F., Enders, A. C.: The fine structure of the guinea pig visceral yolk sac placenta. Amer. J. Anat. **127**, 397–414 (1970b)

King, B. F., Enders, A. C.: Protein absorption and transport by the guinea pig visceral yolk sac placenta. Amer. J. Anat. **129**, 261–288 (1970c)

King, B. F., Enders, A. C.: Protein absorption by the guinea pig choriollantoic placenta. Amer. J. Anat. **130**, 409–430 (1971)

King, B. F., Tibbitts, F. D.: The fine structure of the chinchilla placenta. Amer. J. Anat. **145**, 33–56 (1976)

Krzyzowska-Gruca, St., T. H. Schiebler: Experimentelle Untersuchungen am Dottersackepithel der Ratte. Z. Zellforsch. **79**, 157–171 (1967)

Kunzel, W., Moll, W.: Uterine O$_2$ consumption and blood flow of the pregnant uterus. Experiments in pregnant guinea pigs. Z. Geburtsh. Perinat. **176**, 108–117 (1972)

Larsen, J. F.: Electron microscopy of the chorioallantoic placenta of the rabbit. I. The placental labyrinth and the multinucleate giant cells of the intermediate zone. J. Ultrastruct. Res. **7**, 535–549 (1963a)

Larsen, J. F.: Histology and fine structure of the avascular and vascular yolk sac placentae and the obplacental giant cells in the rabbit. Amer. J. Anat. **112**, 269–284 (1963b)

Laulanie, F.: Sur le processus vaso-formatif qui preside a l'edification de la zone fonctionelle du placenta maternel dans le cobaye. C. R. Soc. Biol. (Paris) **3**, 506–509 (1886)

Leichtweiß, H.-P., Schröder, H.: Untersuchungen uber den Glukosetransport durch die isolierte, beiderseits künstlich perfundierte Meerschweinchenplacenta. Pflugers Arch. ges. Physiol. **325**, 139–148 (1971)

Leichtweiß, H.–P., Schröder, H.: Personal communication 1976

Leissring, J. C., Anderson, J. W.: The transfer of serum proteins from mother to young in the guinea pig. I. Prenatal rates and routes. Amer. J. Anat. **109**, 149–155 (1961)

Levitz, M., Condon, G. P., Money, W., Dancis, J.: The relative transfer of estrogens and their sulfates across the guinea pig placenta: Sulfurylation of Estrogens by the placenta. J. biol. Chem. **235**, 949–954 (1960)

Luft, J. H., Wood, R. L.: The extraction of tissue protein during and after fixation with osmium tetroxide in various buffer systems. J. Cel. Biol. **19**, 46A (1963)

Manning, P., Inglis, R., Green, S., Fishman, W.: Characterization of placental alkaline phosphatase from the rabbit, guinea pig, mouse and hamster. Enzymologia **39**, 307–318 (1970)

Maunsbach, A. B.: The influence of different fixatives and fixation methods on the ultrastructure of rat kidney proximal tubule cells. I. Comparison of different perfusion fixation methods and of glutaraldehyde, formaldehyde and osmium tetroxide fixatives. J. Ultrastruct. Res. **15**, 242–282 (1966a)

Maunsbach, A. B.: Effects of varying osmolality, ionic strength, buffer systems and fixative concentration of glutaraldehyde solutions. J. Ultrastruct. Res. **15**, 283–309 (1966b)

Minot, C. S.: Uterus and Embryo. I. Rabbit, II. Man. J. Morph. **2**, 341–460 (1889)

Moll, W., Künzel, W.: Blood pressures in the uterine vascular system of anaesthetized pregnant guinea pigs. Pflügers Arch. **330**, 310–322 (1971)

Moll, W., Künzel, W.: The blood pressure in arteries entering the placentae of guinea pigs, rats, rabbits, and sheep. Pflügers Arch. ges. Physiol. **338**, 125–131 (1973)

Moll, W., Künzel, W., Ross, H.–G.: Gas exchange of the pregnant uterus of anaesthetized and unanaesthetized guinea pigs. Resp. Physiol. **8**, 303–310 (1970)

Money, W. L.; Dancis, J.: Technique for the in situ study of placental transport in the pregnant guinea pig. Amer. J. Obst. Gynec. **80**, 209–214 (1960)

Mossman, H. W.: The rabbit placenta and the problem of placental transmission. Amer. J. Anat. **37**, 433–497 (1926)

Mossman, H. W.: The comparative morphogenesis of the foetal membranes and accessory uterine structures. Contr. Embryol. Carneg. Instn. **26**, 129–247 (1937)

Mossman, H. W.: In: "Gestation" (ed. C. A. Villee) New York: Josiah Macy Jr., Foundation, 1959

Müller, G., Fischer, W. M.: Über den fetalen und maternen Blutkreislauf in der Meerschweinchen-plazenta. Anat. Anz., Erg. Heft. 121, 231–239 (1968)

Padykula, H.: A histochemical and quantitative study of enzymes in the rat's placenta. J. Anat. (Lond.) 92, 118–129 (1958)

Padykula, H. A., Deren, J. J., Wilson, T. H.: Development of structure and function in the mama-lian yolk sac. I. Developmental morphology and vitamin B_{12} uptake of the rat yolk sac. Develop. Biol. 13, 311–348 (1966)

Palade, G. E.: A study of fixation for electron microscopy. J. exp. Med. 95, 285–297 (1952)

Pease, D. C.: Histological techniques for electron microscopy. Sec. edition. New York and London: Academic Press 15–81, 1960

Perrotta, C. A.: Fetal membranes of the canadian porcupine, erethizon dorsatum. Amer. J. Anat. 104, 35–39 (1959)

Perrotta, C. A., Lewis, P. R.: A histochemical study of placental esterases in the guinea pig and in the human. J. Anat. (Lond.) 92, 110–117 (1958)

Petry, G., Kühnel, W.: Histotopographische und cytologische Studien an den Embryonalhüllen des Meerschweinchens. Z. Zellforsch. 59, 625–662 (1963)

Petry, G., Kuhnel, W.: Der Feinbau des Dottersackepithels und dessen Beziehung zur Eiweißresorp-tion (Kaninchen). Z. Zellforsch. 65, 27–46 (1965)

Piziak, V. K., Gavienowski, A. M.: Biosynthesis of progesterone from endogenous steroid precursor by the guinea-pig-placenta. J. Reprod. Fertil. 32, 283–285 (1973)

Ponse, K., Weihs, D., Libert, O., Dovaz, R.: Trophoblastomes ovariens et leur activite endocrine chez le cobaye. Acta endocr. (Kbh.) 17, 355–365 (1954)

Pytler, R., Strasser, H.: Die Vorgänge im Meerschweinchenuterus von der Inokulation bis zur Bildung des Plazentardiskus. Z. Ges. Anat. 76, 386–420 (1925)

Read, M.: The intra-uterine growth-cycles of the guinea pig. Arch. Entwickl.-Mech. Org. 35, 708–724 (1913)

Reale, E., Pipino, G.: La distribuzione della succinodeidrogenasi nella placenta di alcuni mammiferi. Studio istochimico. Arch. ital. Anat. Embriol. 64, 318–356 (1959)

Reichert, C. B.: Über die Bildung der hinfälligen Haute der Gebärmutter und deren Verhältnisse zur Placenta uterina. Mullers Arch. 1848

Reichert, C. B.: Beitrage zur Entwicklungsgeschichte des Meerscheinchens. Abh. Akad. Wiss. Berlin 1861

Reichert, C. B.: Entwicklungsgeschichte des Meerschweinchens. Abh. Akad. Wiss. Berlin 1862

Reynolds, L., Young, M.: The transfer of free α-amino nitrogen across the placental membrane in the guinea pig. J. Physiol. (Paris) 214, 583–597 (1971)

Roberts, C. M., Perry, J. S.: Histricomorph embryology. Symp. zool. Soc. (London) 34, 333–360 (1974)

Rood, J. P., Weir, J.: Reproduction in female wild guinea pigs. J. Reprod. Fertil. 23, 393–409 (1970)

Sabatini, D. D., Bensch, K. G., Barnett, R. J.: New means of fixation for electron microscopy and histochemistry. Anat. Rec. 142, 274 (1962)

Sabatini, D. D., Bensch, K. G., Barnett, R. J.: The preservation of cellular ultrastructure and enzymatic activity by aldehyde fixation J. Cell. Biol. 17, 19–58 (1963)

Sax, G., Kaufmann, P.: Enzymhistochemie der Meerschweinchen-Subplacenta. Acta. Anat. in press (1977)

Schiebler, T. H., Knoop, A.: Histochemische und elektronenmikroskopische Untersuchungen an der Rattenplacenta. Z. Zellforsch. 50, 494–552 (1959)

Schiebler, T. H., Kaufmann, P.: Über die Gliederung der menschlichen Placenta. Z. Zellforsch. 102, 242–265 (1969)

Schiechl, H.: Der Chemismus der OsO_4-Fixierung und sein Einfluß auf die Zellstruktur. Acta Histo-chem. Suppl. X, 165–171 (1971)

Schneider, S., Kaufmann, P.: Einflüsse des Fixationsmodus auf die Ultrastruktur der Rattenniere. (in prep. 1977)

Schniewind, H., Asshauer, E.: Die Hämodynamik des fetalen und maternen Placentakreislaufes des Meerschweinchens. Arch. Gynäk. 198, 93–99 (1962)

Schröder, H.: Hämodynamik und Stofftransport in der Placenta, Untersuchungen an der isolierten Meerschweinchenplacenta. Habilitationsschrift, Hamburg 1975

Schröder, H., Leichtweiß, H.-P., Madee, W.: The transport of D-glucose, L-glucose and D-mannose across the isolated guinea pig placenta. Pflügers Arch. ges. Physiol. 356, 267–275 (1975)

Schröder, H., Stolp, W., Leichtweiß, H.-P.: Measurements of Na$^+$-transport in the isolated, artificially perfused guinea pig placenta. Amer. J. Obstet. Gynec. 114, 51–58 (1972)

Seelig, H.-P., Römheld, R.: Untersuchungen zur histochemischen Lokalisation der Leucin- und Cystinaminopeptidase (Oxytocinase) in der Placenta. Histochem. 18, 30–39 (1969)

Selenka: Studien zur Entwicklungsgeschichte der Tiere. Die Keimblattumkehr im Ei der Nagetiere. Wiesbaden 1884

Shepherd, J. T., Whelan, R. F.: The blood flow in the umbilical cord of the fetal guinea pig. J. Physiol. (Paris) 115, 150–157 (1951)

Simmer, H. H.: Placental hormones. In: Biology of gestation (ed. N. S. Assali) Vol. I. The maternal organism. P. 290–354. New York, London:Academic Press 1968

Smith R. E., Farquhar, M. G.: Lysosome function in the regulation of the secretory process in cells of the anterior pituitary gland. J. Cell. Biol. 31, 319–347 (1966)

Spee, G. F. v.: Beitrag zur Entwicklungsgeschichte der frühen Stadien des Meerschweinchens bis zur Vollendung der Keimblase. Arch. Anat. Entwickl. Gesch. 44–60 (1883)

Spee, G. F. v.: Vorgange bei der Implantation des Meerschweinchens in die Uteruswand. Verh. anat. Ges. (Jena) 131–136 (1896)

Spee, G. F. v.: Die Implantation des Meerschweinchens in die Uteruswand. Z. Morph. Antrop. 3, 130–182 (1901)

Spiegel, A.: Versuchstiere. Stuttgart: Gustav Fischer Verlag, 1976

Stara, J., Nelson, N., Hoar, R.: [131] J-uptake by pregnant guinea pigs an their fetusses during early embryogenesis. Anat. Rec. 160, 433 (1968)

Starck, D.: Embryologie. 3. Aufl. Stuttgart: Georg Thieme Verlag, 1975

Stark, J., Kaufmann, P.: Semidünnschnitt-cytochemische Untersuchungen zum Transport und Stoffwechsel von Kohlenhydraten in Placentazotten. Verh. anat. Ges. (Jena) 67, 251–256 (1973)

Stark, J., Kaufmann, P.: Infarktgenese in der Placenta. Arch. Gynak. 217, 189–208 (1974)

Stelter, U., Kaufmann, P.: Morphometrie der Meerschweinchenplacenta (in prep. 1977)

Tahmisian, T. N.: Use of the freezing point method to adjust the tonicity of fixing solutions. J. Ultrastruct. Res. 10, 182–188 (1964)

Thomsen, K., Schniewind, H., Schultze-Mosgau, H., Kramer, M., Jost, U.: Technik und Hamodynamik der beidseitigen Perfusion der Meerschweinchenplacenta in vivo. Arch. Gynak. 203, 226–233 (1966)

Thorn, W., Kaufmann, P., Muldener, B.: Kohlenhydratumsatz, Energiedefizit und Plasmapolypenbildung in der Placenta nach Vergiftung mit Monojodacetat und NaF. Arch. Gynak. 216, 175–183 (1974)

Thorn, W., Kaufmann, P., Muldener, B., Freese, U.: Einfluß von 2,4-Dinitrophenol, Monojodacetat, Natriumfluorid und Hypoxie auf Plasmapolypenbildung in der Placenta von Meerschweinchen Arch. Gynak. 221, 203–210 (1976)

Twardock, A. R.: Placental transfer of calcium and strontium in the guinea pig. Amer. J. Physiol. 213, 837–842 (1967)

Uhlendorf, B., Kaufmann, P.: Die Entwicklung des Placentastieles beim Meerschweinchen (in prep. 1977

Vollrath, L.: Das Enzymmuster der Meerschweinchenplacenta und seine Veranderungen im Verlauf der Schwangerschaft. Histochem. 4, 397–419 (1965)

Welsch, F.: Choline acetyltransferase in aneural tissue: Evidence for the presence of the enzyme in the placenta of the guinea pig and other species. Amer. J. Obstet. Gynec. 118, 849–856 (1974)

Wilde, W. S., Cowie, B., Flexner, L. B.: Permeability of the placenta of the guinea pig to inorganic phosphate and its relation to fetal growth. Amer. J. Physiol. 147, 360–369 (1946)

Wislocki, G. B., Deane, H. W., Dempsey, E. W.: Histochemistry of the rodents placenta. Amer. J. Anat. 78, 281–346 (1946)

Wislocki, G. B., Dempsey, E. W.: Histochemical age-changes in normal and pathological placental villi (hydatiform mole, exlampsia). Endocrinology 38, 90–109 (1946a)

Wislocki, G. B., Dempsey, E. W.: Histochemical reactions in the placenta of the cat. Amer. J. Anat. 78, 1–46 (1946b)

Wislocki, G. B., Dempsey, E. W.: Histochemical reactions of the placenta of the pig. Amer. J. Anat. 78, 181–226 (1946c)

Wislocki, G. B., Padykuly, H. A.: Reichert's membrane and the yolk sac of the rat investigated by histochemical means. Amer. J. Anat. **92**, 117–151 (1953)

Wolfer, J., Kaufmann, P.: Die Ultrastruktur der Meerschweinchen-Subplacenta. (in prep. 1977)

Wong, C. T., Morgan, E. H.: Placental transfer of iron in the guinea pig. Quart, J. exp. Physiol. **58**, 47–58 (1973)

Wynn, R. M.: Morphology of the placenta. In: Biology of gestation (ed. N. S. Assali), Vol. I. The maternal organism. P. 93–184. New York, London: Academic Press 1968

9. Subject Index

Other Reviews of Interest in this Series

Part 5: **Gossrau, R.**: Die Lysosomen des Darmepithels. 74 figures. 95 pages. 1975. ISBN 3-540-07271-3

Part 6: **Thorn, L.**: Die Entwicklung des Cortischen Organs beim Meerschweinchen. 23 figures. 97 pages. 1975. ISBN 3-540-07301-9

Volume 52

Part 1: **Ibrahim, M. Z. M.**: Glycogen and its Related Enzymes of Metabolism in the Central Nervous System. 13 figures. 89 pages. 1975. ISBN 3-540-07454-6

Part 2: **Cau, P.; Michel-Béchet, M.; Fayet, G.**: Morphogenesis of Thyroid Follicles in Vitro. 16 figures. 66 pages. 1976. ISBN 3-540-07654-9

Part 3: **Tiedemann, K.**: The Mesonephros of Cat and Sheep. Comparative Morphological and Histochemical Studies. 47 figures. 119 pages. 1976. ISBN 3-540-07779-0

Part 4: **Haug, F.-M. Š.**: Sulphide Silver Pattern and Cytoarchitectonics of Parahippocampal Areas in the Rat. Special Reference to the Subdivision of Area Entorhinalis (Area 28) and its Demarcation from the Pyriform Cortex. 49 figures. 73 pages. 1976. ISBN 3-540-07850-9

Part 5: **Phillips, I. R.**: The Embryology of the Common Marmoset (Callithrix jacchus). 22 figures. 47 pages. 1976. ISBN 3-540-07955-6

Part 6: **Nobiling, G.**: Die Biomechanik des Kieferapparates beim Stierkopfhai. 25 figures. 52 pages. 1977. ISBN 3-540-08038-4

Volume 53

Part 1: **Baur, R.**: Morphometry of the Placental Exchange Area. 37 figures. 65 pages. 1977. ISBN 3-540-08159-3

Springer-Verlag Berlin · Heidelberg · New York